W9-BNT-228

PRINTED CIRCUIT ASSEMBLY DESIGN

Printed Circuit Assembly Design

Leonard Marks

James Caterina

McGraw-Hill
New York San Francisco Washington, D.C. Auckland Bogotá
Caracas Lisbon London Madrid Mexico City Milan
Montreal New Delhi San Juan Singapore
Sydney Tokyo Toronto

Library of Congress Cataloging-in-Publication Data

Marks, Leonard, date.
    Printed circuit assembly design / Leonard Marks, James Caterina.
      p.   cm.
    ISBN 0-07-041107-7
    1. Printed circuits—Design and construction.　I. Caterina, James.　II. Title.

TK7868.P7 M38   2000
621.3815'31—dc21   00-032906

McGraw-Hill

A Division of The McGraw-Hill Companies

2 3 4 5 6 7 8 9 0 DOC/DOC 0 5 4 3 2 1

ISBN 0-07-041107-7

The sponsoring editor for this book was Stephen S. Chapman, the editing supervisor was David E. Fogarty, and the production supervisor was Pamela A. Pelton. It was set in Vendome ICG by Pat Caruso of McGraw-Hill's Professional Book Group composition unit, Hightstown, N.J.

Printed and bound by R. R. Donnelley & Sons Company.

This book is printed on recycled, acid-free paper containing a minimum of 50% recycled, de-inked fiber.

CONTENTS

V

PREFACE

Printed circuit assemblies (PCAs) are key building blocks found in almost every commodity being made today that has any electronics content. Requirements for small, light, inexpensive, high-performance products are driving the complexity of these designs to astronomically high levels. Printed circuit design is no longer a "seat of the pants" activity but rather a highly sophisticated technical effort. The present span of processes and technologies associated with the design of PCAs is very broad. The amount of detail connected with these processes and technologies is enormous!

The primary intent of this book is to describe the procedures and considerations that are involved in successfully producing PCA designs. It provides up-to-date information on the processes, methods, and tools used to design state-of-the-art assemblies. The book discusses the considerations and requirements that influence how PCAs are configured and details the typical steps in the design process and the ways they are related to each other. It also explains the different approaches and tools commonly used to develop and document designs and suggests effective approaches for organizing and managing these activities.

This book is structured as a detailed process roadmap for a broad spectrum of readers involved in PCA design, from the neophyte just starting out to the experienced designer, and it also can be used as guidance for engineers and managers who deal in some way with this technology. There are many other books and reference materials available that address the subject of printed circuit design. Some cover particular aspects of design in great detail (e.g., high-speed circuitry, surface-mount high-density interconnections, etc.), and some discuss design in much less detail as a part of larger books that primarily describe printed circuit fabrication, assembly, and test processes. Most contain a wealth of data and reference information contained in charts, tables, curves, and mathematical formulas that can be used to help make design decisions. This book does not offer such detailed information, although it will alert readers when it is needed and usually will point out where it can be found. Instead, the material found in the following chapters hopefully will illustrate and explain the complexities and nuances of the design process and guide and support all those who are responsible for some aspect of the development, fabrication, assembly, or testing of PCAs.

—Leonard Marks

—James Caterina

CHAPTER 1

Introduction

Many factors have contributed to the evolution and widespread application of printed circuit technology as we know it today. Undoubtedly, the most significant of these are

- Development of a repeatable and relatively economical process for manufacturing circuit boards

- Invention of the solid-state transistor and then incorporation of many of these semiconductor devices into a single, integrated circuit

- Use of the eutectic soldering process to produce highly reliable electrical connections between the circuit board and its associated electronic components

Serious use of printed circuit assemblies (PCAs) began shortly before the end of World War II. The boards were made initially by attaching a layer of copper foil to a paper-based laminate, applying a silk-screened ink pattern to the foil, and etching it to form an interconnecting circuit. One of the first applications for these crude assemblies was in a radio set. Another type of circuit board, designed for use in proximity fuses, was made by directly printing a metallic conductive pattern on an insulator. By the end of the war, these were being produced successfully in very large quantities.

The processes, materials, and equipment that evolved from the latter approach eventually became the foundation for today's hybrid circuit technology. The former circuit board fabrication concept, augmented by development of processes for electroless metallization, electroplating, photo-imageable resist, precision drilling, multilayer lamination, along with the availability of a large variety of copper-clad dielectric materials, matured into what is now the most widely used technique in the electronics industry for implementing circuit interconnections.

The growth of printed circuit technology has paralleled the growth of the electronics industry. There are very few electronics products, or products with electronics content, presently being manufactured that do not contain one or more PCAs. Circuit boards have progressed from being simply a substitute for hard-wired interconnections to becoming an integral functional part of today's complex, high-performance circuitry. Because of this, issues that designers of these assemblies must now deal with have expanded well beyond simply placing components on a board and finding paths to interconnect them.

1.1 Printed Circuit Assembly Applications

Designers should have an appreciation for the broad variety of systems and products that employ PCAs. They also need to understand how these systems and products will be used and the environments in which they will function.

An effective design will result in a PCA that performs effectively and reliably throughout the product's life cycle, which encompasses fabrication, assembly, test, storage, transportation, and operation. During its life, a PCA may be required to survive exposure to a wide range of environmental conditions, such as temperature extremes, humidity, mechanical shock and vibration, atmospheric variations, harmful chemicals, and electromagnetic radiation.

Other major design influences and constraints related to a PCA's application include

- Manufacturing cost
- Time to market (schedule)
- Access for testing, adjustment, replacement, repair
- Size and weight
- Reliability
- Availability of parts and materials

1.1.1 Common Types of PCAs

Printed circuit assemblies are classified in a variety of ways:

- By type of circuitry used (digital, analog, mixed analog-digital radio frequency, microwave)
- By types of electronic components used and how they are mounted (through-hole, surface-mount, mixed-technology, components mounted on one side or both sides)

- By board construction (single-sided, double-sided, multilayer, flex, rigid-flex, stripline)
- By application and performance required (general electronic products, high-performance, dedicated service products, high-reliability products)
- By design complexity, circuit density, and manufacturability (low, moderate, high)

Most PCAs incorporate electronic components and interconnections that perform specific circuit functions. Some, however simply serve as system-level interconnections, such as connector motherboards, or flex-circuit wiring harnesses. Printed circuit technology is also being used increasingly to produce integrated electronic components such as multichip modules (MCMs) and microwave integrated circuit (MIC) modules. Figure 1.1 shows examples of some of the more common types of PCAs.

1.1.2 Where and How PCAs Are Used

Standards developed by the Institute for Interconnecting and Packaging Electronic Circuits (IPC) identifies the following three classes of electronic products in which PCAs are used:

- *Class 1.* General products, including consumer items, computers and peripherals, and some military systems
- *Class 2.* Dedicated service products, including communication equipment, business machinery, industrial controls, instruments, and military systems, where extended life and reliable service is needed
- *Class 3.* High-reliability products, including equipment and systems where continuous performance or performance on demand is essential

The basic intended functions of a printed circuit board are to support and interconnect electronic components. For most applications, these functions are required to be performed reliably, consistently, and as cost-effectively as possible. As mentioned earlier, circuit boards have evolved from being simple, passive parts of an electronic assembly to becoming an active circuit element whose design has a major effect on the performance of a product.

Flex

Analog

Digital

Figure 1.1
Examples of common types of PCAs.

PCAs serve as key hardware building blocks for products found in all segments of the electronics market. The two largest areas of application (in terms of hardware cost/sale value) are communications equipment and computer systems. Other significant PCA utilization product areas include industrial controls, instrumentation, automotive, consumer, military/space, and business/retail. A substantial increase in the capabilities and application of digital circuitry [e.g., microprocessors, analog-to-digital (A/D) converters, and digital signal processors] has resulted in an explosion of new electronic products using this technology. This has considerably expanded the need for complex, high-performance PCAs for use in games, mobile phones, cameras, video recorders, TVs, pagers, CD players, global positioners, and many other products. In addition, functions previously performed by mechanical or electro-mechanical methods are rapidly being converted to pure electronic implementations. Two examples are appliance controllers and under-hood automotive sensors and controllers.

1.2   Basic PCA Elements

PCAs are commonly composed of a similar set of basic elements. These are

- Passive and active electronic components that perform the PCA's intended circuit functions
- A printed circuit board that supports the components and provides interconnections between them
- One or more connectors that are the electrical interface between the PCA and the rest of the system/product
- Mechanical parts for mounting parts and hardware to the board, attaching the PCA to the system/product, providing thermal paths, and supporting and stiffening the assembly

Figure 1.2 identifies some of these basic elements.

1.2.1   Passive and Active Components

Passive components are parts that exhibit a fixed or controlled value and perform an elementary function in a circuit. Exam-

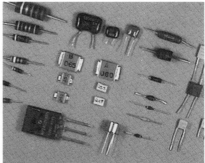
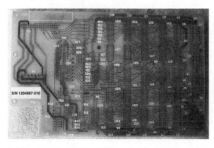
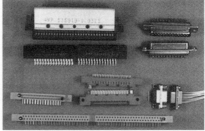

Figure 1.2
Typical PCA elements.

ples are resistors, capacitors, inductors, and conductors. They are available packaged as discrete devices with leads for through-hole mounting or in chip packages for surface mounting. Multiple passive parts (mainly resistors) also may be enclosed in a single package. Figure 1.3 shows some typical passive components.

Active components, such as diodes, transistors, silicon-controlled rectifiers (SCRs), and integrated circuits, provide variable parametric values to a circuit and can perform specific, high-level activities by functioning, for example, as an amplifier, a switch, a rectifier, a signal detector, or a one-way conductor. Active devices are available as discrete leaded or chip parts, but the vast majority are provided as integrated circuits (ICs), which contain a large (mostly very large!) number of active components. These ICs may be packaged for either through-hole or surface mounting and are able to perform a variety of complex circuit functions. ICs are designed for either general-purpose or application-specific use. General-purpose components, such as digital logic circuits, memory arrays, and operational amplifiers, can be used as building

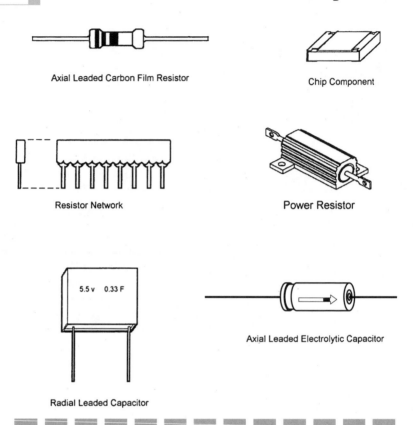

Axial Leaded Carbon Film Resistor

Chip Component

Resistor Network

Power Resistor

5.5 v   0.33 F

Radial Leaded Capacitor

Axial Leaded Electrolytic Capacitor

Figure 1.3
Typical passive parts (no scale).

blocks to construct a variety of functional circuits. Application-specific ICs such as microprocessors, A/D converters, or digital signal processors are usually designed for a single purpose or to perform an explicit complex function. Figure 1.4 shows some typical active components.

As the preceding figures show, components are packaged in many different physical configurations. The two main packaging families are through-hole leaded devices and surface-mounted parts. Input-output lines (I/Os) for surface-mounted packages may be provided as leads, plated tabs, or pin or ball grid arrays. Devices are also classified by the type of package material used (i.e., plastic, ceramic, metal), allowable operational environment (i.e., temperature, humidity), critical parameter tolerances, and projected reliability (i.e., operational life).

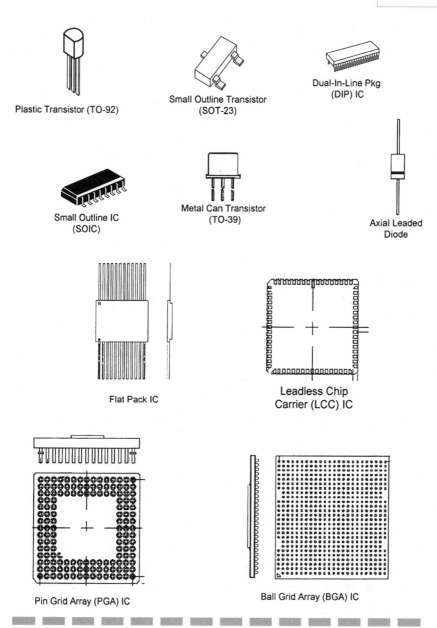

Plastic Transistor (TO-92)

Small Outline Transistor
(SOT-23)

Dual-In-Line Pkg
(DIP) IC

Small Outline IC
(SOIC)

Metal Can Transistor
(TO-39)

Axial Leaded
Diode

Flat Pack IC

Leadless Chip
Carrier (LCC) IC

Pin Grid Array (PGA) IC

Ball Grid Array (BGA) IC

Figure 1.4
Typical semiconductor devices (no scale).

1.2.2 Interconnections

A printed circuit board, which supports and interconnects the electronic components, is composed of two basic elements. One part is the base substrate, which is usually a combination of an insulating dielectric and a reinforcing material that is embedded in the dielectric. The other constituent is the metal foil from which conductors are formed to produce the circuit paths between the components.

A number of dielectric materials are used for printed circuit boards, some with reinforcement and some without. Selection usually is based on electrical properties, compatibility with the operating environment, processing constraints, and cost limitations. Epoxy-based resins and polyimide resins are by far the most frequently used materials.

The predominant reinforcement used in circuit board laminates is cloth that is woven from glass, quartz, or Kevlar fibers. Glass cloth is least costly and easiest to process. Recently, more exotic reinforcing materials have become available for use where special physical or electrical properties are needed, such as random-fiber (nonwoven) mat, ceramics, carbon fibers, metal cores, and others.

Finished conductors can consist of multiple materials. Those on external circuit layers are primarily copper, frequently covered with a thin layer of tin/lead (solder) plating. In selected areas of a board, gold or some other low-oxidizing plated material may cover the conductors instead. Conductors on inner circuit layers usually are not plated. One or both of the copper conductor surfaces on a laminate may be treated by oxidizing them to enhance adhesion to the insulating substrate. The thickness of conductors on a board is determined by a weight rating in ounces per square foot of the copper foil. Copper used on most circuit boards range from 2 to $\frac{1}{2}$ oz (1 oz is approximately 0.0014 in thick). The odd way the weight is specified can be traced back to when copper was used as a roofing material.

In many of today's high-performance PCAs, the physical and electrical characteristics of the substrate and conductor materials may have an effect on how the assembly performs its intended function. In addition, material selection can determine the producibility of a design.

1.2.3 Connectors

Connectors are used to provide a mechanical and electrical interface between a PCA and the rest of the world. As opposed to hard-wired methods for establishing such an interface, connectors furnish the ability to rapidly plug a PCA into its next-higher assembly or system and easily unplug it, if required. Two basic types of connectors are currently used in PCAs:

- *Edge-card connectors* use an edge of a circuit board as the male portion, with etched and plated contacts on the surface of the board that plug into a female connector mounted on a motherboard or a chassis (Fig. 1.5).

- A *two-part connector* consists of a set of male or female contacts contained in an insulating body that is mounted on a circuit board. The board connector mates with a matching connector (having the opposite pin type) mounted on a motherboard or a chassis. Two-part connectors are available that are designed for soldering to a board using either through-hole or surface-mount technology (Fig. 1.6).

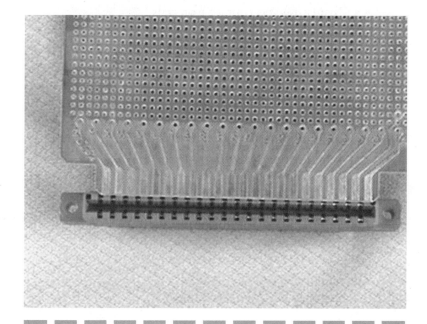

Figure 1.5
Edge-card connector and mating circuit board.

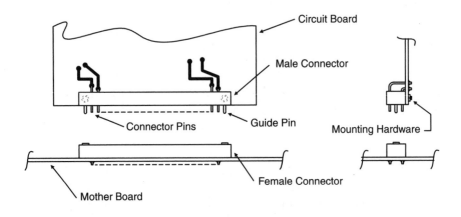

Figure 1.6
Two-part connector.

Connectors are available in a large variety of configurations and many different combinations of physical and electrical properties, including

- Quantity and spacing of contacts
- Insulating and contact materials
- Mechanical mounting method
- Operating temperature
- Insertion/withdrawal forces
- Keying capability
- Electrical characteristics
  - Contact resistance
  - Insulation resistance
  - Current rating
  - Operating voltage

Circuit board connectors are also characterized by their application. Application categories include

- Commercial
- Industrial
- Military
- Space

Connectors intended for commercial or industrial use usually will have been tested and certified by the Underwriter's

Laboratory (UL). Connectors used in military or space systems will have been tested and qualified to a set of stringent electrical and environmental requirements.

1.2.4 Mechanical Parts

Different kinds of mechanical parts and hardware may be included in a PCA design. They may perform many functions, including

■ Aiding in assembly insertion and withdrawal (handles, ejectors)

■ Strengthening the circuit board and preventing bow and twist (brackets)

■ Mounting large, heavy components (clamps, spring clips)

■ Providing thermal paths (heat sinks, edge clamps)

■ Forming conductive elements (eyelets, rivets, terminals)

Mechanical parts should be placed to maximize ease of installation, since manual procedures are usually required for mounting them. This is particularly important for surface-mounted assemblies where the majority of components are installed with automated equipment. Mechanical parts should be selected and positioned so that they can be attached either before or after the components are mounted and not interrupt the automated assembly process. Also, adequate electrical and physical clearance should be provided between these parts and circuit elements, taking into account dimensional tolerances associated with their location and package dimensions.

1.3 PCA Design Process Life Cycle

It is important for printed circuit designers to understand that layout is just a part of the overall PCA design process, although one that is particularly crucial to the successful development of a design. Since PCAs form the basic building blocks of most electronic systems, the design process actually begins by defining the functions that each of these building blocks must perform. A design has little value until it is turned into operating, production-configuration hardware and integrated into a system or product. When integration has been implemented successfully, the design process is considered to be complete.

All the interrelated activities that occur within the boundaries defined by these two events define a PCA's design process life cycle. It is generally recognized that the PCA design effort frequently paces a program's overall hardware development schedule and constitutes a major part of its nonrecurring (engineering and design) cost. Although it is not as well recognized, the actual board layout activity is not usually the largest cost or schedule element of a PCA design cycle. It has been the experience of both authors (based on a combined total of 60 years involved with designing PCAs) that the costs associated with board layout activities are about 15 to 20 percent of the total PCA cost of design. Schedule time required for layout is about 12 to 15 percent of the total. The pie charts in Fig. 1.7 graphically show the cost and schedule relationships that usually exist between the circuit design, layout, and integration activities needed to produce a new PCA design.

1.3.1 Scope of the Design Process

As shown in the process flowchart in Fig. 1.8, PCA design activities begin with the functional and physical partitioning of a system. Partitioning is a critical process step because the decisions made at

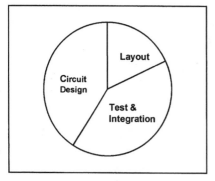
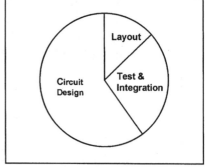

Nonrecurring (design) Cost   Schedule

Figure 1.7
Cost and schedule relationships: circuit design/layout/test and integration.

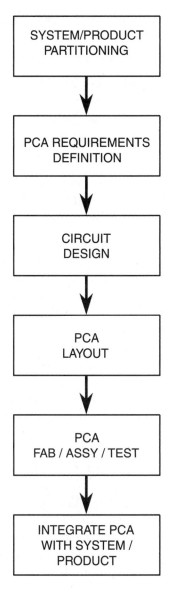

Figure 1.8
PCA design life cycle.

SYSTEM/PRODUCT PARTITIONING

↓

PCA REQUIREMENTS DEFINITION

↓

CIRCUIT DESIGN

↓

PCA LAYOUT

↓

PCA FAB / ASSY / TEST

↓

INTEGRATE PCA WITH SYSTEM / PRODUCT

this point will affect all downstream design and production activities. Partitioning entails

- Identifying the major functions of a system/product and how they are linked to each other
- Selecting the circuit technologies and component types required to implement the functions

- Grouping the functions together that are most closely related to each other

- Estimating the amount of circuitry required for each group

- Allocating the circuitry to individual PCAs based on available circuit board area, as defined by physical constraints

- Confirming that the circuitry will fit as allocated and the PCA will be producible by performing a test placement and doing a preliminary interconnection capability analysis

- Performing a preliminary physical design analysis (thermal, vibration/shock) to ensure that the PCA will function properly in its intended environment

After circuitry has been allocated to an assembly, the specific operational requirement of that assembly should be identified, parametrically defined, and documented. This set of physical and circuit specifications becomes the basis for defining the PCA's design characteristics and will be used as the criteria for verifying its performance.

A detailed circuit design is then developed, resulting in the creation of a schematic diagram that identifies all electronic components required to implement the PCA's functions and describes how they are to be interconnected. The schematic also specifies the circuit interfaces between the PCA and the rest of the system/product. Breadboarding or modeling and simulation may be performed on all or portions of the design to predict how the circuitry will perform over its intended range of operating conditions. This activity also may serve to identify critical portions of the circuitry that will require special attention during physical layout of the printed board.

The schematic, along with a set of layout rules, becomes the source of the data used to prepare a board layout. Prior to starting this activity, a library containing information about the parts included in the design should be in place. For manually produced layouts, a physical description of each part type, defining its package outline and pin locations, is sufficient. Library data required for layouts generated by computer-aided design (CAD) is more extensive and should describe a component's

- Package type
- Reference designator type (R, C, CR, U, etc.)

- Mounting technology
- Package dimensions
- Dimensional origin location
- Package I/O location and size dimensions
- I/O numbering and functions
- Circuit characteristics of each I/O
- Thermal and weight characteristics

After a layout has been completed, it should be checked for completeness, correctness, and adherence to all applicable design rules. Drawings, data files, and phototools required for manufacturing the design are then prepared, and a small quantity of first article assemblies may be produced and tested. If possible, these development models or prototypes should be made using the same processes, equipment, and tooling that will be employed in building production quantities of the PCA. In this way, the design and its manufacturing data can be validated before a commitment is made to build it in large quantities.

The last part of the design process involves integrating the PCA into its system/product. The basic purpose of this activity is to verify that the design will perform its intended functions in an operating environment. This can be done in various ways, and Table 1.1 shows some typical approaches used for integration testing along with advantages and disadvantages of each.

1.4 The Team Approach to PCA Design

Designing a PCA is a team effort that requires the active participation of representatives from diverse disciplines. The levels of participation of the various disciplines will vary during the design cycle, depending on the activity under way. An effective team does more than transfer data between disciplines; it encourages all members to participate in making design trade-offs and decisions so that the responsibility for the end result is shared by everyone on the team.

TABLE 1.1

Printed Circuit Assembly System Test/Integration Methods

Strategies for PCA Integration	Typical Applications	Advantages (+)/Disadvantages (−)
Operational system (sometimes called a test bed or hot mockup)	■ Digital circuit designs for new systems ■ Radio frequency circuit redesigns for existing systems ■ When at-speed testing is required	+ Short test development schedule + Pre-debugged + Potentially low cost + Test software (SW) also can be used for operational system SW and built-in-test (BIT) SW + Can be combined with overall system integration effort − Must build and test sequentially − Hardware limited (one test bed)
Dedicated PCA test station (or one-time benchtop setup)	■ Radio frequency circuit designs for new systems ■ Large, complex systems ■ Temperature-sensitive circuitry ■ When at-speed testing is required	+ Test hardware (HW) and SW can be developed in parallel + Test station also can be used for production testing − Test HW and SW debugging effort is required − Unique test software is required − Costly to design and build test station
Universal in-circuit tester	■ Mainly used to identify non-functional parts and board faults ■ Digital circuitry ■ Manufacturing processes used are mature/repeatable/automated	+ Minimal hardware design required + Test SW generally supported by development tools, i.e., CAD/CAE import of library and simulation vectors − High up-front cost − Interface adapter performance degrades the operating environment (vs. test bed)
Built-in-test (BIT)	■ Enhances Operational System and Dedicated Station approach	+ Based on HW architecture + Reusable SW + Vertical testing + At-speed test + Can replace some functional test HW and SW + Facilitates integration − Software intensive (cost and schedule considerations)

1.4.1 Design Team Structure

A PCA design team usually has an overall team leader and includes participants representing the different disciplines that are involved in the development of the design. During each phase of the design cycle, one or two disciplines may be tasked with the primary responsibility for implementing that part of the process. The team leader may be from a company's program management organization or from technical management. Technical disciplines and job classifications that are part of a PCA design team and participate in varying degrees in making design decisions may include

- Systems engineering
- Circuit engineering
- Mechanical engineering
- PCA layout
- Software engineering
- Test engineering
- Manufacturing engineering
- Data management
- Quality assurance
- Maintenance and field support
- Purchasing
- Key subcontractors

1.4.2 Team Responsibilities

As PCA design activities proceed through the various stages shown in Fig. 1.8, lead responsibilities change. Table 1.2 lists the tasks that typically are performed in each phase of the design process and shows which disciplines usually have lead responsibility for completing each task and which disciplines are primary participants or contributors.

1.5 Industry Standards, Organizations, and Associations

The establishment of industry-wide standards for PCA design, materials, board fabrication, assembly, and quality is the result of cooperation and general agreement among the printed circuit design and manufacturing community. Acceptance and applica-

TABLE 1.2

Functional Design Activity

Phase	Task	Disciplines (' = Lead)
System/product partitioning	1. Allocate PCA functions to be implemented in either HW or SW and characterize function interfaces. 　　Perform trade studies as needed.	Project/system engineering (') Circuit design SW engineering
	2. Partition system/product to PCA level.	Project/system engineering (') Circuit design Mechanical engineering SW engineering Layout design
	3. Develop PCA block diagram showing structure and signal/data flow within each function. 　　Perform electrical/mechanical density check and preliminary thermal/structural analysis.	Project/system engineering (') Circuit design Mechanical engineering SW engineering
	4. Define manufacturing/test approach for PCA.	Manufacturing/test engineering (')
PCA requirements definition	1. Define and document requirements (incl. PCA interfaces).	Project/system engineering (') Circuit design Mechanical engineering SW engineering
	2. Establish design rules (circuit design, mechanical design, layout, board fabrication, assembly, circuit testing).	Project/system engineering (') Circuit design Mechanical engineering SW engineering Layout design Manufacturing/test engineering
Circuit design	1. Decompose functional blocks into a detailed circuit, design. 　　Perform circuit simulation/analysis. 　　Create additional schematic library symbols (if needed). 　　Perform fit check. 　　Perform preliminary thermal/structural analysis.	Circuit design (') Layout design Mechanical engineering
	2. Select component types and values. 　　Perform selection risk analysis for availability, potential obsolescence, and single-source vulnerability.	Circuit design (') Components engineering Purchasing
	3. Establish PCA schematic and drawing/database and transfer to layout function.	Circuit design (') Layout design
PCA layout	1. Define PCA-unique physical/circuit layout requirements and restrictions. 　　Create additional physical part library shapes (if needed).	Circuit design Layout design (') Mechanical engineering
	2. Place parts and test points, perform test route, and finalize placement. 　　Perform final thermal/structural analysis.	Circuit design Layout design (') Mechanical engineering Manufacturing/test engineering
	3. Route PWB interconnections.	Layout design (')

TABLE 1.2

Functional Design Activity (Continued)

Phase	Task	Disciplines (' = Lead)
PCA layout	4. Verify that all design requirements are met.	Circuit design Layout design (') Mechanical engineering Manufacturing/test engineering
	5. Prepare manufacturing data, tools, and drawings.	Layout design(') Drafting
Fabrication/ assembly/test	1. Fabricate/inspect/test initial bare boards. Develop/document/prepare unique programming and tools/fixtures for bare board fabrication and test.	Manufacturing/test engineering (')
	2. Assemble/inspect/test initial PCAs. Develop/document/prepare unique programming and tools/fixtures for PCA assembly and test.	Manufacturing/test engineering (')
Integration	1. Perform integration and tests at next higher levels of assembly. Develop/document/prepare unique programming and tools/fixtures for PCA integration and test	Project/system engineering (') SW engineering Manufacturing/test engineering

tion of these standards enable original equipment manufacturers (OEMs), component and material suppliers, fabricators and assemblers, and product customers to communicate with each other using commonly understood terminology and criteria.

There are a number of organizations and associations that work with the printed circuit industry, some directly and some peripherally, to develop, document, and disseminate standards. Foremost among these is IPC—Association Connecting Electronics Industries. Other organizations that generate standards applicable to the printed circuit industry include the American National Standards Institute (ANSI), the U.S. government (DoD, NASA), Underwriter's Laboratory (UL), the Federal Communications Commission (FCC), and the International Organization for Standardization (ISO).

1.5.1 IPC—Association Connecting Electronics Industries

IPC is a source of specifications, standards, methods, and guidelines that are specifically applicable to PCA design and manufacturing. These materials are available in printed form, on CD-ROM, and in VCR video format. Some also can be downloaded from the IPC Web site.

Among the areas covered by IPC publications are

- Design standards for rigid and flexible circuit boards
- Performance requirements
- Design guides
- Surface-mount standards
- Board material specifications and requirements
- Multilayer construction
- Preparation and transfer of manufacturing data
- Guidelines and requirements for electrical testing
- Manufacturing processes and procedures
- Circuit board and assembly quality standards

1.5.2   American National Standards Institute (ANSI)

ANSI facilitates development of standards by establishing consensus among qualified groups and ANSI-accredited developers that supports the establishment of national and international standards. ANSI publications that are most applicable to printed circuits are those which define drafting procedures and documentation formats.

1.5.3   U.S. Government

Until recently, PCAs intended for use in military or space systems had to be designed and built in accordance with a variety of government-issued standards and specifications. PCA requirements were mainly defined by military specifications that established the layout rules, material standards, manufacturing processes, and quality levels to be met by every design.

The U.S. government, mainly the Department of Defense (DoD), has now moved away from imposing these specifications on new designs, and in most cases, industry standards are taking their place. Many military specifications are no longer being updated to reflect changes in printed circuit technology and are no longer being required for new electronic programs.

1.5.4 Underwriter's Laboratory (UL)

UL is an independent, not-for-profit product safety testing and certification organization. It defines operational requirements for various types of electronic products and tests them to ensure that they will function safely. Products UL tests that incorporate PCAs may include computers, network electronics, medical devices, and telecommunications equipment. UL also performs electromagnetic interference/electromagnetic compatibility (EMI/EMC) testing to assess conformance to FCC regulations.

1.5.5 International Organization for Standardization (ISO)

ISO 9000 is a generic term for a set of standards created by the ISO that define the requirements for a quality system to be established by manufacturing and service companies. It is applicable to the design and manufacture of PCAs in that it describes a system for ensuring the quality of the output of any process.

Unlike a product standard, ISO 9000 is not specifically applicable to a particular commodity. It addresses a facility's quality system, including the methods, practices, and techniques that are integrated with established design and manufacturing processes to ensure that the output of the processes satisfies customer requirements.

2

Key Factors Influencing Design

There are many important, interrelated factors, considerations, and requirements that influence how printed circuit assemblies (PCAs) are configured. These include, but are not limited to

- *Application considerations.* How the PCA will be used in the end product or system

- *Performance requirements.* What the PCA must achieve, functionally

- *Interfaces.* The relationship and interactivity within the PCA's final installation

- *Manufacturing and test compatibility.* The ability to produce the design as reliable, cost-effective hardware

- *Cost and schedule constraints.* Programmatic requirements associated with budgetary and delivery issues

This chapter serves as a guide to the various parts of this book by listing the sections that specifically address and discuss each of the factors that have a key role in determining how a PCA can be optimally configured.

2.1 Application Considerations

These include

- Use and maintenance of the end product
- The PCA's operating environment
- Size and weight constraints
- Accessibility, safety, and reliability requirements
- Design standards

2.1.1 Use and Maintenance of the End Product

Sections of the book that discuss *PCA use, repair,* and *maintenance* include 3.1.1, 3.4, 3.7.3, 4.4.9, 5.5.1, 6.1.3, and 6.7.4.

2.1.2   PCA's Operating Environment

Sections of the book that deal with *environmental concerns* include 1.1, 1.2.1, 1.2.2, 1.2.3, 1.3.1, 3.1, 3.1.3, 3.2.3, 3.2.6, 3.3, 3.3.2, 3.3.3, 3.3.4, 3.4.3, 3.4.4, 3.4.6, 3.5.1, 3.5.6, 3.5.7, 3.6.3, 3.6.4, 3.6.5, 3.6.6, 3.7, 3.7.4, 4.1, 4.1.3, 4.3.6, 4.4, 4.4.8, 4.5, 4.6.5, 5.4, 5.4.4, 5.5, 5.5.2, and 7.3.4.

2.1.3   Size and Weight Constraints

Sections of the book that address *size and weight constraints,* include 1.1, 3.1.3, 3.3.1, and 5.3.3.

2.1.4   Accessibility, Safety, and Reliability Requirements

Sections of the book that consider *safety and reliability* include 1.1, 1.1.1, 1.1.2, 1.2.1, 1.5.4, 3.1.3, 3.2.6, 3.3, 3.3.4, 3.4.2, 3.4.4, 3.6, 3.6.2, 3.6.6, 3.7.1, 3.7.4, 4.1.3, 4.3.5, 4.3.6, 4.4, 4.4.1, 4.4.2, 4.4.6, 4.4.8, 4.4.9, 4.7.9, 4.8, 5.3.1, 5.4.1, 5.4.3, 5.4.4, 5.5, 5.5.1, 5.5.2, 5.5.3, 5.6, 5.6.2, and 7.2.3.

2.1.5   Design Standards

Sections of the book that deal with *design standards* include 1.1.2, 1.5, 3.2.3, 3.2.6, 3.4.3, 3.6.4, 4.1.10, 4.2.1, 4.2.2, 4.2.4, 4.2.5, 4.3.3, 4.6.5, 5.4.4, 6.1.1, 6.1.3, 6.2.3, 6.4.7, 7.1.4, 7.1.5, 8.1.3, and 8.1.4.

2.2   Performance Requirements

2.2.1   Types of Circuitry Used

Sections of the book that discuss *types of circuitry* include 1.1.1, 1.1.2, 3.2.3, 3.2.5, 4.2, 4.4.3, 4.4.4, 4.5.1, 4.7.7, and 5.3.5.

2.2.2   Circuit Frequency/Clock Speed

Sections of the book that deal with *circuit frequency and clock speed* include 3.2.6, 3.6.2, 4.4.4, 4.5.1, 4.6.2, 4.6.3, 4.6.4, 4.6.5, 4.7.5, 4.7.7, 5.3.5, and 5.4.2.

2.2.3   EMI/EMC

Sections of the book that cover *EMI/EMC considerations* include 1.5.4, 3.1.3, 3.2.3, 3.2.6, 3.3.2, 3.4.4, 4.4.2, 4.4.3, 4.4.4, 4.6.3, and 4.6.5.

2.2.4   Component Types

Sections of the book that address *component types* include 1.1.1, 1.2.1, 3.2.2, 3.2.3, 3.2.5, 3.2.6, 3.4.3, 3.4.4, 3.5.1, 3.6.2, 3.7.3, 4.1.6, 4.2.5, 4.3.1, 4.3.2, 4.3.3, 4.4.7, 4.7.9, 5.3.1, and 5.3.3.

2.2.5   Board Material and Construction

Sections of the book that discuss *board material and construction* include 1.1.1, 1.2.2, 1.5.1, 3.2.6, 3.3.1, 3.3.3, 3.3.4, 3.4.2, 3.4.4, 3.4.9, 3.5.1, 3.5.3, 3.5.4, 3.5.5, 3.6.2, 4.1.7, 4.1.10, 4.4.8, 4.6.1, 4.6.4, 4.7.4, 4.7.8, 4.7.10, 5.3.2, 5.3.3, 5.3.5, 5.4.1, 5.5.2, 6.3.1, and 6.3.2.

2.2.6   Heat Removal

Sections of the book that deal with *heat removal* include 1.2.4, 3.1.3, 3.3.1, 3.3.2, 3.3.3, 3.4.7, 3.6.2, 3.6.5, 3.6.6, 4.3.2, 4.3.6, 4.4.1, 4.4.8, 4.6.1, 4.7.9, 5.3.3, 5.4.4, 5.5.1, and 5.5.2.

2.2.7   Shock and Vibration Protection

Sections of the book that discuss *shock and vibration protection* include 1.3.1, 3.3.2, 3.4.4, 3.6.5, 3.6.6, 3.7.4, 4.1.3, 4.3.6, 4.4.1, 4.4.8, 4.5.3, 5.4.4, and 5.5.1.

2.3 Interfaces

2.3.1 External Circuit and Power Connections

Sections of the book that deal with *external circuit and power con-nections* include 1.2.3, 1.3.1, 3.1.3, 3.2.1, 3.2.2, 3.2.3, 3.3.4, 3.4.4, 4.1.1, 4.2.3, 4.4.1, and 4.4.7.

2.3.2 Mechanical Mounting

Sections of the book that address *mechanical mounting* include 1.2.3, 1.2.4, 3.2.6, 3.3.2, 3.4.4, 3.6.5, and 4.4.2.

2.3.3 Heat-Transfer Paths

Sections of the book that discuss *heat-transfer paths* include 1.2.4, 3.3.2, 3.3.3, 3.6.5, 3.6.6, 4.1.1, 4.4.2, 4.4.8, 4.6.1, and 5.3.1.

2.3.4 Test and Adjustment Access

Sections of the book that cover *test and adjustment access* include 3.1.3, 3.2.6, 3.3.4, 3.6.2, 3.6.6, 3.7.1, 3.7.2, 4.1.1, 4.1.3, 4.3.5, 4.3.6, 4.4.1, 4.4.6, 4.7.9, 4.7.10, 5.4.2, 5.4.3, 6.5.2, 6.5.3, and 6.5.4.

2.3.5 PCA Removal and Replacement

Sections of the book that deal with *PCA removal and replacement* include 3.1.3, 3.3.4, 3.3.5, 3.4.7, 3.6.5, and 4.1.3.

2.3.6 Software Interfaces

Sections of the book that describe *software considerations* include 3.1.2, 3.2.1, and 3.2.2.

2.4 Manufacturing and Test

2.4.1 Industry Board Fabrication Capabilities

Sections of the book that address *industry board fabrication capabilities* include 3.4.9, 3.5.1, 3.5.3, 4.6.4, 5.3.2, 5.3.3, 5.3.5, 6.1.1, 6.1.2, and 8.2.3.

2.4.2 Industry Assembly and Test Capabilities

Sections of the book that address *industry assembly and test capabilities* include 1.5.4, 3.2.6, 3.4.9, 3.6.2, 4.4.9, 5.3.1, 5.4.2, and 8.2.4.

2.4.3 Part and Material Availability

Sections of the book that deal with *part and material availability* include 1.2.1, 1.2.2, 1.2.3, 3.2.3, 3.2.5, 3.3.1, 3.5.1, 3.6.1, 3.6.2, 4.1.7, 4.1.10, 4.7.4, 4.7.10, 5.3.2, 5.3.3, 5.3.5, and 5.5.2.

2.5 Cost and Schedule Constraints

2.5.1 Cost Limitations

Sections of the book that address *cost and schedule constraints* include 1.1, 1.2.2, 1.3, 3.1, 3.1.1, 3.2.5, 3.4, 3.4.2, 3.4.4, 3.5.1, 3.5.3, 3.5.7, 3.6, 3.6.2, 4.1.8, 4.1.9, 4.1.10, 4.3.5, 4.4.5, 4.4.9, 4.6.1, 4.7.10, 5.1.2, 5.1.3, 5.3, 5.3.1, 5.3.2, 5.3.3, 5.3.4, 6.1.2, 6.5.2, 6.7.3, 7.1.1, 7.1.2, 7.2.1, 7.2.2, 8.1.4, 8.1.6, 8.1.8, 8.2.1, 8.2.3, and 8.2.5.

2.5.2 Schedule Limitations

Sections of the book that address *schedule limitations* include 1.1, 1.3, 3.1, 3.1.1, 3.2.5, 3.4, 3.4.4, 3.5.1, 4.4.9, 4.7.10, 5.1.2, 6.1.2, 6.7.3, 7.1.1, 7.1.2, 7.2.1, 7.2.2, 8.1.4, 8.1.6, 8.1.8, 8.2.1, 8.2.3, and 8.2.5.

Design Process Flow

The process flow diagram in Fig. 3.1 shows the normal sequence of activities in the printed circuit assembly (PCA) design process and their relationship to each other. It displays the general chronology of the design effort, but it does not show the iteration and feedback paths inherent in a typical design effort.

Ideally, each step in the PCA design process should be accomplished in a logical sequence, and there should be as few "loopbacks," or repeated activities, as possible. Since this is not a perfect world, such an ideal is rarely achieved. Therefore, in order to maximize the cost and schedule efficiency of a PCA design effort, review checkpoints, or *gateways*, should be established at critical points in the process. Successful completion of a review can minimize the risk that some part of the design activities preceding the review was done improperly. The process flow diagram in Fig. 3.1 indicates the points in the process where design reviews are most effective.

It is important that organizational functions that are directly or indirectly involved in the life cycle of a PCA should be participants in these reviews. Any or all of the following could be included:

■ The project engineer responsible for the overall PCA design

■ The PCA circuit design engineer

■ The board layout designer

■ The engineer responsible for integrating the PCA into the next level of assembly

■ Production engineering

■ Test engineering

■ Components engineering

■ Engineers responsible for designing other PCAs that will interface with the one being designed

In a large company, each of these functions could be a separate entity, whereas in smaller companies, some of them may be performed by the same organization or even by the same person. Design reviews are discussed in more detail in Chap. 5.

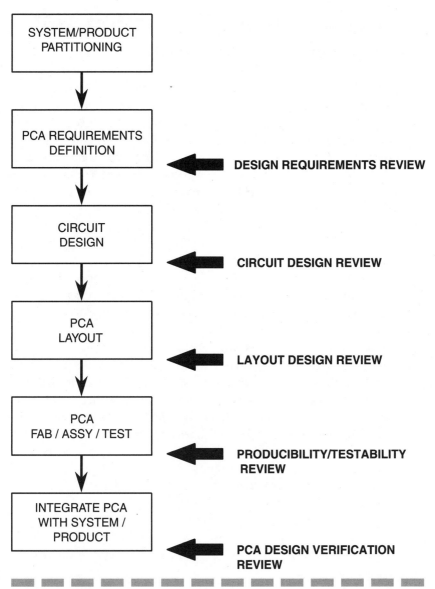

Figure 3.1
PCA design reviews.

3.1 Requirements Allocation and Functional Partitioning

The first steps in defining the configuration of a new PCA are taken long before detailed circuit design and layout activities begin. The PCA design process actually begins with the definition of system- or product-level functional elements based on what the system or product is supposed to do. For example, if the product is to be a new desktop computer, the functional elements required may include a processor, memory, data storage, a power source, an input capability, a display, interfaces to external devices, management of all the functions, and a way for them to communicate with each other. This defines its architectural structure.

Once a system's functional requirements are specified, decisions are made about how to implement them, based on such constraints and requirements as available technology, cost, schedule, risk, manufacturing and test capabilities, operating environment, maintainability approaches, and other considerations. The design approach to be used to implement each functional element is based on these decisions. In the world of electronics, either hardware or software, and sometimes a combination of both, can implement most functions. Those functional elements allocated to hardware can be realized in a variety of ways, one being by circuitry on a PCA. Figure 3.2 is a graphic representation of how the functional elements of a system or product are allocated to an implementation approach.

3.1.1 Definition of Functional Performance Requirements

Performance requirements are based on what the end product must do, how it is to do it, and under what conditions it will be expected to fulfill its intent. Top-level functional requirements are the "whats." From these are derived the "hows," which identify the potential methods of implementation for each function. At the system level, a function can be treated initially as a

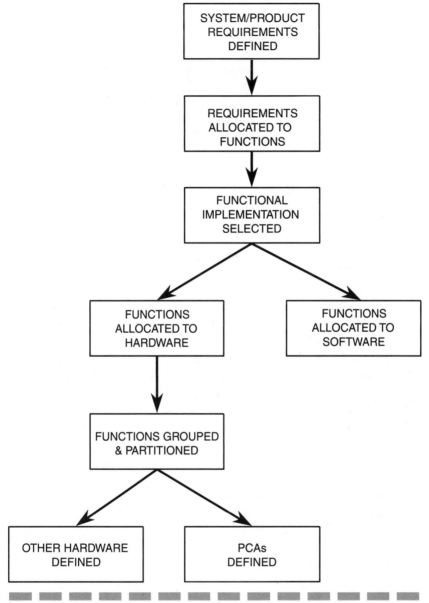

Figure 3.2
Allocation of functional elements of a system/product.

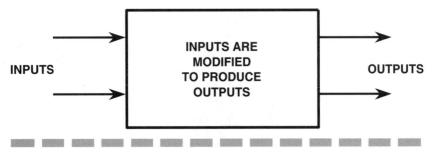

Figure 3.3
Definition of a function.

"black box" with inputs. Activities take place within the box that modify the inputs and produce a set of outputs. Figure 3.3 is a graphical representation of a generic function.

Since different ways of realizing a function are usually possible, a comparison of the advantages and disadvantages of each approach should be conducted. Tradeoff studies are usually performed to evaluate the relative merits of alternative concepts and technical approaches. Data obtained from a trade study can supply the basis from which the best design approach can be selected considering the requirements, performance, cost, schedule, process, risks, and resource constraints.

3.1.2   Hardware/Software Implementation

One of the most significant decisions to be made early in the design process is whether a system or subsystem element is to be implemented in hardware, software, or a combination of both. Very few electronic functions are now performed solely with software or hardware. Most require a combination of both to meet their requirements.

Software controls the electronic pulses digital circuitry needs to operate. Analog circuitry senses and processes signals based on physical values, such as temperature, pressure, or distance, but these data usually are converted to digital information somewhere in a system. The same is true for radio frequency (RF) circuitry. Many software functions are performed directly by preprogrammed physical devices, such as programmable read-only memory (PROM) or field-programmable gate array (FPGA) devices. Programming circuit functions in software is attractive because software is much easier to modify than hardware once a system has been built.

3.1.3 PCA Partitioning

A number of physical and performance requirements and constraints should be considered when partitioning circuit functions and allocating them to individual PCAs. These include circuit performance, physical constraints, producibility and testability, and maintainability.

1. Circuit performance considerations may involve
   - Length of signal connections
   - Segregation of digital and analog/RF circuitry
   - Power distribution
   - EMI/EMC
   - Component power dissipation

2. Physical constraints may include
   - Size and shape of the space allocated for PCAs
   - Area needed for components and interconnections
   - Quantity of inputs and outputs required
   - Environmental conditions
   - Heat removal

3. Producibility and testability needs may include
   - Circuit density (components and interconnections)
   - Ability to identify and isolate circuit faults
   - Component packaging and mounting
   - Characteristic tolerances (circuit parameters and physical features)

4. Maintainability requirements may have to be accommodated, including
   - Ease of PCA removal and replacement
   - Access for adjustment and test (in place and out of the product)
   - Provision of features for repair and modification

3.2 Circuit Design

Once functions are assigned to a PCA and the technical approaches to be used to implement these functions are selected, detailed circuit design activity is begun. The initial step in this process is to identify the interoperability of the functions (basically, how they work with each other) and determine how signals are to flow between them and to and from functions external to the PCA. The operational requirements for these interfaces are then established, in terms of compatibility, data content and structure, physical connections, and signal timing, if appropriate. The next step is to define how these functions are to be physically accomplished by selecting part types, logic families, and use of custom-designed or off-the-shelf parts. With these decisions made, the circuitry required for each function on the PCA can be designed in detail.

As the detailed circuit design proceeds, it may be desirable for the circuit engineer to breadboard or construct a simulation model for each function as it is developed to demonstrate that the design is conceptually sound. When circuitry is completed for all the functions on the assembly, it may be appropriate to invest in a breadboard or a simulation model of the entire PCA to validate the design and reduce potential risk that the finished hardware may not meet all its requirements.

Figure 3.4 diagrams the flow of this process, portions of which are described in more detail in the following subsections.

3.2.1 Development of Circuit Functional Flow and Theory of Operation

How functions on a PCA are linked to each other, and to other parts of the system electronically, physically, and with software, is frequently described by a hardware/software functional flow block diagram. This may be accompanied by a documented explanation of the functional theory of operation that describes the basis for arriving at that particular detailed design configuration.

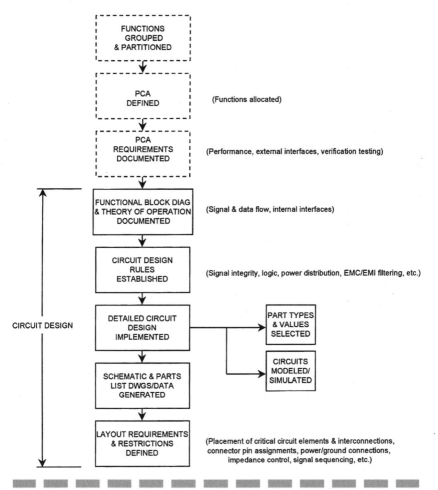

Figure 3.4
Circuit design process flow.

A functional flow diagram shows the relationships of critical hardware and software entities on the PCA. It describes the internal interfaces that are required for system/product operation. It also may identify software program sequences that configure and/or activate the various hardware functions and describes the hardware's response(s) to these program sequences, including timing and dwell issues and nonlinear operating conditions. Figure 3.5 shows a portion of a simple functional flow diagram.

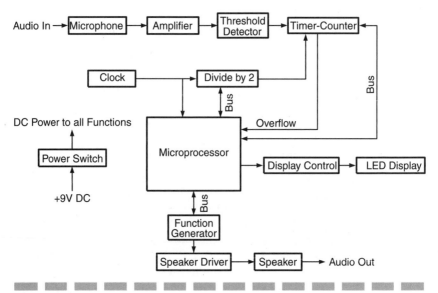

Figure 3.5
An example of a functional flow diagram.

3.2.2   Definition of Internal Hardware/Software Interfaces

An important part of the functional design decision-making process is choosing how data are to be handled and stored on a PCA. Options include use of off-the-shelf static or dynamic random access memory (RAM) devices, prewired read-only memory (ROM), or programmable memory devices called *PROMs.*

A programmable component of one or more types is usually procured and then programmed to provide a unique function prior to installing it on a PCA. They are not easily reprogrammed, and thus the program is usually identified as *firmware.* As part of the detailed design process, a programming specification is prepared for the firmware for each device. Figure 3.6 shows a portion of a typical PCA schematic containing programmable read-only memory. Figure 3.7 shows an example of an EPROM device. The use of memory devices and the type selected depend on what data are to be processed, their origin, and their destination.

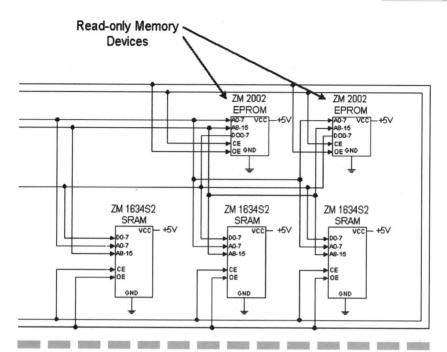

Figure 3.6
Portion of a PCA schematic containing read-only memory devices.

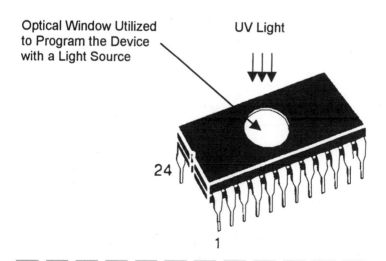

Figure 3.7
Example of an EPROM.

3.2.3 Selection of Part and Logic Families

Circuit performance requirements generally dictate the choice of a part-type or logic-type family of components. Industry standards have been established defining the key characteristics and parameters for specific device families so that they can be obtained from more than one manufacturer and still be compatible with each other.

Digital logic families are defined by their electrical operating parameters and signal-transfer characteristics. Analog devices are selected based on functionality and performance characteristics (e.g., signal response, isolation, environmental sensitivity, EMC/EMI, operating voltage, ground integrity, and noise). Table 3.1 lists typical characteristics of common logic families. A key

TABLE 3.1

Typical Characteristics of Common Logic Families

Family	Propagation Delay (ns)	Power Dissipation (mW)	Comments
Std TTL-7400	10	10	Original version, nearly obsolete
Schottky TTL-74S00	3	20	Schottky transistors, faster that standard TTL
Low-power Schottky TTL-74LS00	9	2	
Advanced low-power Schottky 74ALS00	4	1	Replaces 74LS00
FAST TTL-74F00	3	4	For high-speed circuits
4000 CMOS-74C/4000	90	1 @ 1 MHz	Poor drive capability
High-speed CMOS-74HC00	10	1.5 @ 1 MHz	Good drive capability
Advanced CMOS-74AC00	3	0–1 @ 1 MHz	For high-speed circuits
Emitter-coupled logic (ECL), 100 K	1	40–50	Fastest, low noise margin

Note: Values are for single-gate performance.

consideration when selecting part and logic families for implementing circuit functions on a PCA is their compatibility with each other and with the circuitry used in other parts of the system that interfaces with the PCA.

3.2.4 Circuit Modeling and Simulation

Circuit simulation can be performed by circuit designers using computer-based tools to confirm the operation of a function without the need to build and test breadboard hardware. These computer programs, which are available commercially, can emulate the operation of circuitry over a specified range of parameters and conditions, considering part types and values, signal parameters, operating conditions, and physical relationships.

Models representing complex circuit functions and devices and logic types must be created and integrated into the simulation software. Many of these models are furnished by the provider of the tools or by the device manufacturers, but in some cases the simulator's user may have to develop the model. This could be a costly and time-consuming effort. Some simulation programs allow actual components to be used in place of models, if none exist.

Simulators can deal with large, complex functions and combinations of functions up to the level of a complete PCA. Simulators also can be used to analyze the testability of a circuit design.

3.2.5 Definition and Design of Custom Parts

Although employment of off-the-shelf components is usually the most cost-effective way of implementing circuit functions, use of custom-designed, application-specific devices is sometimes a more logical choice. This may be the case when standard components will not provide the required performance or board real estate is limited and the custom part is much smaller. Custom devices also may be used if the nonrecurring cost of design and the recurring cost of the part and the manufacturing and test are less than those required for equivalent standard parts.

Significant cost, schedule, and performance risk may be associated with choosing to incorporate a custom-designed part in a circuit. Not until it is designed, fabricated, tested, and integrated into a PCA circuit can the designer be completely certain that the custom part will function as required. If it does not, a whole new round of design, build, test, and integration activities may be needed, and this could have a devastating impact on a design project.

Many types of custom electronic components are available, including

- Fully custom application-specific integrated circuits (ASICs)
- Standard-cell ASICs
- Gate-array ASICs
- Programmable logic devices (PLDs), such as field-programmable gate arrays (FPGAs)
- Thin- or thick-film hybrids
- Chip-scale packages (CSPs)

Table 3.2 summarizes the key features of each type.

TABLE 3.2

Key Features of Various Types of Custom Devices

Type	Key Features
Fully custom integrated circuit (ASIC)	Developed for a specific application, utilizing unique circuit elements on an IC chip; most costly and time-consuming to develop.
Standard-cell ASIC	Developed for a specific application by combining and interconnecting predesigned circuit elements, taken from a standard cell library; less development cost and time required.
Gate-array ASIC	Developed for a specific application by applying a custom trace pattern that interconnects a predefined array of gates on a chip; least development cost and time required.
Programmable logic device (PLD), such as field-programmable gate arrays (FPGA)	Developed for a specific application by applying a program to a prepackaged device that establishes interconnection links between predefined circuit elements; most costly to produce, but minimal development required.
Thin- or thick-film hybrids	Developed for a specific application by defining and interconnecting passive elements (resistors, capacitors, etc.) on a substrate and attaching active chip devices to complete the circuit; primarily used for analog and stripline circuits.

3.2.6 Development and Application of Circuit and Layout Design Rules

General rules and guidelines to be used when designing a family or group of similar PCAs for a system or product should be established at the program or project level before any design activities begin. These rules can then be tailored (modified) as needed to be consistent with the design requirements for each individual assembly. Design rules generally are derived from industry standards, internal or subcontractor manufacturing and test capabilities and experience, customer requirements, performance specifications, preferred material and component lists, and design tool availability. They typically standardize

- Component types and logic families (specific part numbers, values, tolerances)
- Circuit design rules (loads, terminations, drivers, power distribution, clock signal distribution, grounding, EMI protection, part derating)
- Placement of parts and circuitry (proximity, signal coupling, orientation)
- Interconnection layout (min/max trace dimensions, parallel conductors, stubbing, impedance, order of signal and ground/power layers)
- Mechanical design (board dimensions, construction, material, mounting hardware, thermal management, environmental protection, identification marking)
- Location of signal and test connectors and key connector pin assignments (power/ground, clock lines)
- Test and maintenance (fault isolation, test/adjustment accessibility)

A typical design rules document for a family of PCAs used in a system is shown in Fig. 3.8.

3.3 Mechanical Design

Mechanical design of a PCA entails defining its physical form and structure. This includes establishing the assembly outline

The following document establishes a set of design rules for a family of PCAs used in a specific system/product:

PCA Design Rules for _____

Mechanical:

> Board Type   CFG
> Maximum Height   0.125" (thickness of board, including solder plus component height)
> Solder Mask   Dry Film
> Hardware   per ___(Specification)___
> Environment   per ___(Specification)___
> Marking   per ___(Specification)___ , mark PCA part number by edge farthest from primary connector so that the dash number can be read when PCA is installed in product

Electrical:

> The following are designated as critical nets: _____
(avoid routing signal traces parallel to these on adjacent layers, keep these traces as short as possible)
> Place decoupling capacitors as close as possible to associated logic devices.
> Locate power and ground planes as follows: (layer location, specific part/circuitry association, split planes, embedded circuitry, termination of planes)
> Function swapping is allowed on the following devices: _____
> Pin swapping is allowed as follows: _____
> Place the following parts in close proximity to each other: _____
> The following parts are to be installed in sockets: _____
> Limit bus loading to one TTL load.
> Buffer all bus signals near the primary connector; avoid long stubs at all cost.
> Provide data bus drive for ____ TTL loads.
> Provide system clock distribution system to reduce clock skew.
> Limit clock loading to 4 devices per driver.
> Buffer or decouple all signals that go to the I/O connector.
> Provide for on board reprogramming of all Flash or EEPROM memory.
> Tie all unused inputs high or low (as appropriate) though a resistor.
> Derate all components per reliability guidelines. Capacitors - derate working voltage to 50 % of rated voltage. Resistors - derate power to 50 % of rated power.
> Design for growth (Provide 10% spare gates distributed throughout the circuitry).
> Provide software reset capability.
> Initialize all hardware to a known state.
> Design for fault tolerance to minimize "hard" failures

Testability

> Testability shall include requirements for bare board, in-circuit, PCA-level functional and system-level functional tests:
> Bring out as many unmuxed but buffered outputs as possible for in-system debugging.
> Provide means to initialize circuits with test equipment.
> Add muxes to observe internal modes
> Add test points for observability.
> Add series pull up/down resistor to unused enables, resets, and sets
> Route all CPU control and interface signals to the test connector for Bus monitoring.
> All I/O ports must be read/write ports.
> Provide a means to disable clocks and permit driving the clocks circuits from external test equipment.

Components:

> Avoid all single source components.
> Standardize components with existing PCAs.
> Use either all surface mount components or all through-hole parts (no mixed technology) .
> Choose components from the preferred parts selection list.
> Plastic Parts may be used provided the parts can operate over the temperature range in the application,
> A standard plastic package may be used if:
       - Test data is available on it
       - If it is compatible with thermal and structural mounting requirements
       - If moisture seal requirement has been verified
> Prohibited materials: beryllium, magnesium, mercury, or other toxic flammable or radioactive materials shall not be utilized.

Figure 3.8
Typical design rules for a family of PCAs.

(i.e., length, width, height), circuit board dimensions and how it is mounted, the method of interfacing it to other circuitry, access for removal/replacement, test and adjustment, heat removal, and environmental protection.

3.3.1   Board Configuration

The primary constraints that directly determine the size and shape of a circuit board are where the assembly is to be installed, how it is to be mounted, and the amount and type of circuitry it will contain. Board construction (i.e., material, number of layers, stackup) is mainly influenced by component and interconnection density and circuit performance requirements.

Once the primary form factor of a PCA is established, other features can be defined and located, including mounting hardware, heat sinks, input-output (I/O) connectors, test points and connectors, adjustable parts, and keep-out areas. Manufacturing requirements (i.e., fabrication, assembly, test), such as size and location of tooling holes and datums, optical alignment marks, dimensional tolerances, complex machining requirements, minimum and maximum dimensions of conductor features, plated hole dimensions, and compatibility with standard material panel sizes, also should be considered. Figure 3.9 contains a portion of a drawing of a PCA showing how typical mechanical features and parts are defined and installed.

3.3.2   Support Structure and Mounting Hardware

In most cases the assembly into which a PCA is to installed has been designed before the layout is started, including the way the PCA is to be mounted, the location of the I/O connectors, alignment features, and the mechanical extraction and replacement method. Specific hardware required to perform these functions usually has been selected and must be accommodated in the layout. If this has not yet been done, the mechanical design of the interfacing structure *must* be established, in detail, before circuit layout activities are begun, or the risk of having to make major changes later is significant.

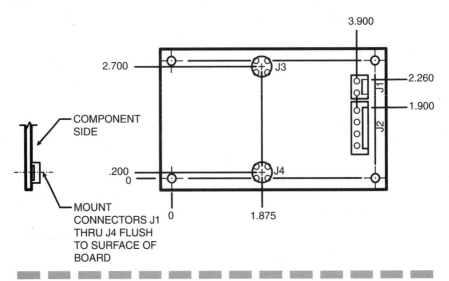

Figure 3.9
Partial PCA assembly drawing showing mounting of connectors.

The support structure for the PCA also has to provide protection against excessive vibration and shock inputs and include adequate heat-removal capabilities. If card stiffeners, special thermal interface hardware, EMI/EMC protection, or other environmentally related design features are needed, they also should be defined before beginning the physical layout. This includes where they are located, how they are attached to the circuit board, and what hardware is used to mount them to the board. Figure 3.10 shows a typical PCA installed in its assembly.

3.3.3 Heat-Sink Design

There are three modes of heat removal from components: conduction, convection, and radiation. Heat sinks are available that use one, two, or a combination of all three, depending on how they are configured and applied. How parts are cooled is a direct function of the PCA's operating environment, how it is mounted in a system, and the internal cooling mechanisms provided by the system itself. The objective of designing a PCA for optimal heat dissipation is to ensure that the maximum allowable

Figure 3.10
PCA installed in a plastic housing.

temperature of components and board material is not exceeded during operation.

If a system's mechanical structure is externally liquid- or air-cooled and the PCA is mounted to the structure, then cooling takes place primarily by conduction of heat to that structure. The designer must make sure that adequate conduction paths exist between hot parts and the heat-exchange surface of the cooled structure.

If internal cooling is provided, either air or liquid, the designer must be aware of flow paths of the coolant and place hot parts on the PCA for maximum exposure to the coolant. If the design configuration or construction of a part limits direct heat removal, then its thermal interface to the coolant must be augmented. Providing an external heat sink for the part, mounting the part on a thermally conductive surface, or attaching it to a board with thermally conductive adhesive can do this. Some examples of heat sinks that can be used to improve a part's heat dissipation are given in Fig. 3.11.

Any supplementary heat-sinking hardware or material used should be compatible with the assembly and cleaning processes used to build the PCA. Entrapment of corrosive or conductive materials must be prevented, either by providing adequate flow paths for cleaning or by sealing the interfaces to preclude entry.

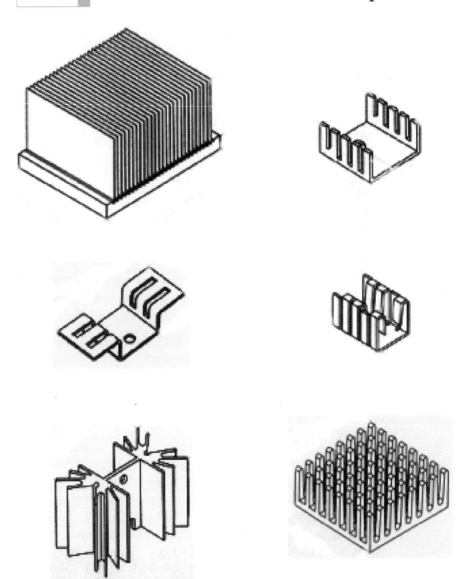

Figure 3.11
Examples of heat sinks for improving thermal dissipation of PCA component parts.

3.3.4   Connectors and Test Points

Connector selection is usually made based on the following factors:

- The quantity and type of signal and power connections required (current, voltage, impedance controlled, shielded)
- The space available for these connections (density)
- How a PCA physically interfaces with the system (plug-in, permanent mount)
- Circuit interfaces with the system (motherboard, hard wiring, flex, directly to adjacent assembly)
- Board construction (single/double-sided, multilayer)
- Type of component mounting (through-hole, surface-mount)
- Environment (temperature, humidity)
- Reliability (number of insertions/withdrawals)
- Application (military, space, industrial, commercial)

Many types of connectors are available combining a wide variety of contact configurations, materials, finishes and spacings, body types and materials, and mounting methods. They generally can be categorized as through-hole (right-angle or straight contact), surface-mount, circular, and edge-card (using the circuit board as the male connector) connectors. Most are soldered to the board, but some may depend on pressure to make contact with the board. In systems where multiple PCAs are installed in the same area, the connectors may be keyed to prevent incorrect placement. Figure 3.12 contains examples of some of the different types of two-part connectors.

Test points or receptacles usually are chosen based on their location and purpose. If test points are required to be accessible while a PCA is mounted in the system, then they are usually soldered to the board's surface near its edge. If a small quantity is needed (8 to 10), individual receptacles can be soldered to the board. They should be sized to be compatible with the test probes that will be used. Larger quantities can be obtained as multireceptacle units and soldered to the board in a single operation. Some PCAs incorporate a separate connector on the board to accommodate attachment of test equipment for complex measurement of operability.

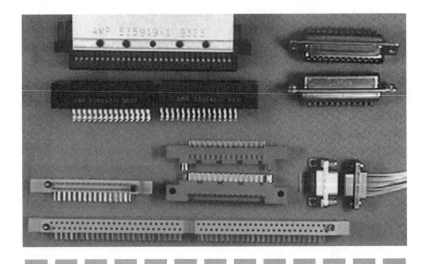

Figure 3.12
Examples of different types of two-part PCA connectors.

3.3.5   Height Restrictions and Keep-Out Areas

There are occasions where the proximity of adjacent structures or other assemblies will restrict the height of components or hardware mounted on a PCA or even prevent parts from being installed in specific areas of the board. These restrictions may be applied to either or both sides of an assembly and must be respected when laying out a board.

Also, consideration must be given to providing clearance for insertion and withdrawal of an assembly, and clearances must be allowed for tooling required for fabrication, assembly, and testing of the PCA. Figure 3.13 is the top view of a PCA showing some typical keep-out areas.

3.4   Board Layout

PCA board layout activity can be implemented using a broad variety of techniques and approaches depending of the type of assembly to be developed, the experience of the designer, the

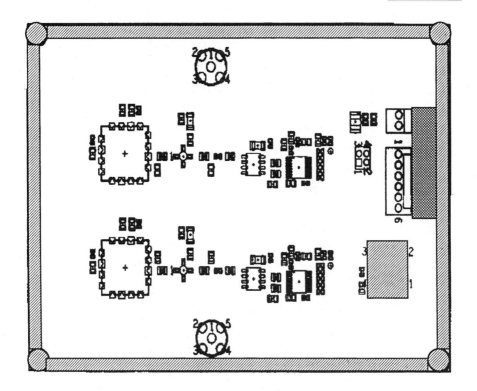

 = Trace / Via Keepout Area (all layers)

= Via Keepout Area (top layer)

Figure 3.13
Typical keep-out areas on a PCA.

tools and resources available, and the fabrication, assembly, and
test processes that will be used to produce it. Normally, the
organization responsible for preparing a design has a prior body
of knowledge to draw on that is based on experience gained
from previous PCA design efforts.

General guidelines are usually in place defining how a PCA is
transitioned from system design, to circuit design, to board lay-

out, to fabrication, assembly, and test, to its integration in the final system. In addition, project-specific design rules should be established and documented specifying

- Circuit and mechanical design data input content and structure (schematic, parts list, thermal/vibration analysis results)
- Layout approach, analyses, restrictions
- Tools and aids to be used
- Board construction options and limitations
- Definition of part types, packages, mounting methods, and footprints
- Layout rules, part placement restrictions, interconnection methodology
- Design reviews and layout verification (checking)
- Documentation requirements
- Generation of manufacturing and test data
- Design data storage and retrieval
- Responsibilities of all organizations involved in the design activity and the interfaces between them
- Cost and schedule requirements and constraints

If any of these basic guidelines and rules are not available to the designer, they should be established before layout activity begins. Starting without clearly defining these requirements creates a serious risk that the end product may not completely or properly perform the function for which it was intended, may not meet its cost targets, and may not be available when it is needed.

3.4.1 Layout Tools

Board layout can be accomplished using either manual methods or computer-based tools. Availability of inexpensive, off-the-shelf, computer-aided design (CAD) programs capable of running on low-cost desktop workstations or personal computers has rendered the manual layout method essentially obsolete. However, situations occasionally arise where a small, simple PCA is needed for a product, and because the organization responsible for producing the design does it so infrequently, there are no automated tools in place.

To produce a PCA layout manually requires a relatively simple set of tools and aids. The basic components needed are transparent, accurately gridded Mylar sheets and scaled cutouts representing parts to be used and showing their leads and land pattern (dolls). Sticky-backed pads and various-width tapes are used to define the conductor patterns. To ensure reasonable precision, the layout should be prepared at a larger scale than the end product (anywhere from 2:1 to 10:1, depending on the dimensions of the board). The basic layout procedure is usually as follows:

1. Draw the board outline, its mechanical features, and keep-out areas to scale on a gridded sheet.

2. Place the part dolls as appropriate on that sheet.

3. Using a separate gridded sheet for each conductor layer, align each with the component sheet and define the location of traces and vias required to complete interconnections between the parts with the tape and pads. This becomes the master artwork set.

4. Manually verify the layout against the schematic, parts list, and mechanical requirement inputs.

5. Photographically reduce the circuit master artwork layers to actual size, and produce the phototools required for manufacturing the bare board.

6. Manually prepare the drawings needed for board fabrication and assembly.

It is also possible to convert the manual layout to computer-readable information after completing step 4 by digitizing the master artwork. Once these data are captured, the rest of the process is similar to the CAD-based layout process.

Use of CAD technology automates much of the printed circuit board layout process. A number of CAD tools are available commercially, and they offer a broad range of capabilities. Many are merged with automated circuit design, analysis, and schematic capture programs, and some offer thermal analysis functions. Most can perform self-check activities and have drafting capabilities for producing manufacturing drawings. All can produce data files in industry-standard format for generation of manu-facturing tools (artwork) and information used by automated manufacturing, assembly, and test equipment. Figure 3.14 shows a screen shot of a CAD system displaying a complex circuit board that has been routed and is being edited to improve its manufacturability.

CAD systems fall into several broad categories based on their capabilities. They are

- The type of platform they run on (usually UNIX-based workstations or personal computers)

- The size and complexity of the boards that can be designed (number of layers, nets, components)

- How much of the layout process is automated (placement, routing, design rule constraint, and analysis)

- The type and variety of outputs that can be produced

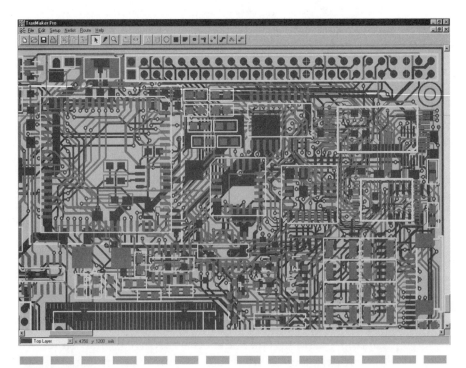

Figure 3.14
CAD system screen shot of layout in progress. (*Used by permission of Protel International, PTY LPD.*)

Like most software, CAD systems continue to expand in capability and sophistication, while their cost keeps going down. As discussed in Chap. 6 on design quality, the use of CAD tools to lay out circuit boards not only lowers the cost of design but also has a significantly positive effect on the quality and performance of the end product.

3.4.2 Board Construction

Before starting the physical layout activity, a designer must determine how the board is to be constructed. This primarily involves

- Defining the board's outline dimensions
- Identifying the types of material to be used; the number, location, and orientation of the signal layers; if ground and power planes are needed; how vias are to be implemented; how components are to be mounted (including connectors and test points); and the location and configuration of mechanical features

In addition to performance and environmental considerations, a PCA's cost target, as well as manufacturability, testability, maintainability, and reliability, must be factored into the selection process. If other PCAs have been or are being designed for a product, these decisions already may have been made. If not, the designer should involve downstream functions (manufacturing, test, field service) in defining the board's construction features.

3.4.3 Definition of Part Outlines and Land Patterns

The definition of a circuit component consists of two elements: its physical description and its functional description. The physical description includes body dimensions and type of material; lead/pin location, dimensions, and material; thermal characteristics; and environmental protection requirements (if any). Also associated with a part is its schematic definition, showing individual internal circuit functions with their related lead/pin designations and the schematic symbology used for the part. The source of this information is data usually provided by the manufacturer of a part in the form of a specification sheet. Figure 3.15 contains a typical vendor specification for an electronic component.

An organization that has responsibility for designing PCAs and does so on a regular basis usually establishes a central component library to collect and maintain the data associated with parts used in the assemblies. This library is normally divided into two sections, one for the physical representation of parts and the other for each part's schematic symbology. As new part types are used, they are added to the library. In this way, circuit engineers can use the schematic symbols in the library as they create designs, and designers can employ the associated physical component descriptions for the layouts. Establishing a PCA parts library has a great many advantages, including standardization and control of component use, consistency between schematic and physical definitions, and maintenance of up-to-date parts information.

Also part of a circuit component physical definition is the description of the land pattern it requires for attachment to a circuit board. The elements of a land-pattern description are

■ The location and shape of the pads

■ For parts with through-the-board leads, the size and location of the holes (plated or unplated)

■ The size and location of related via holes and pads (especially for surface-mounted parts)

■ The size and location of related test points

A land-pattern design standard is derived from the size and shape of a part, the location and configuration of its leads, how it is mounted and electrically attached to a circuit board, the construction of the board, and the processes used for assembly and test. Industry standards generally are used by designers and accepted by PCA manufacturers for establishing land patterns required for common part types. A popular source for land-pattern design guidelines is standards published by IPC—Association Connecting Electronics Industries. They are IPC-2221, "Generic Standard on Printed Board Design," IPC-2222, "Sectional Design Standard for Rigid Organic Printed Boards,"

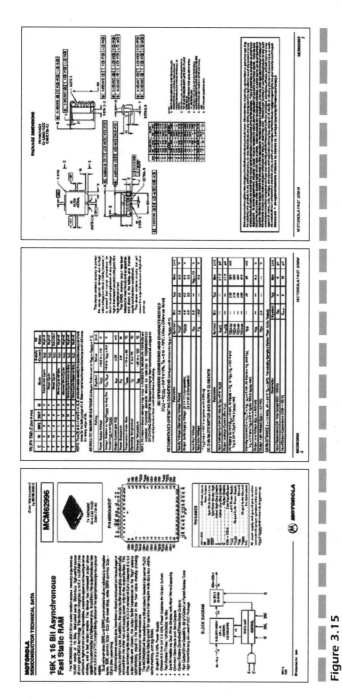

Figure 3.15

Typical component specification sheet (partial). (*Copyright of Motorola, used by permission.*)

and IPC-SM-782, "Surface Mount Design and Land Pattern Standard."

3.4.4   Layout Design Rules

A designer should have a well-defined and documented set of design rules and guidelines in hand before starting the physical layout effort. These rules generally are derived from a number of preestablished requirements and conditions, including

- Circuit performance
- Types of components included in the circuit
- The system operating environment
- Manufacturing and test processes
- Reliability and maintainability
- The circuit board construction and materials
- Cost and schedule limitations

Layout rules are driven primarily by circuit performance requirements and generally are defined by the circuit designer. Constraints typically can be placed on such things as

- Physical grouping together of logical circuit elements
- Proximity or separation of functional groups
- Location of specific parts and circuitry near board connectors
- Location and length of critical interconnections
- Sequencing of connection points in a circuit net
- Configuration of power and ground connections
- Control of circuit parameters of specific connections (impedance, resistance)
- EMI/EMC restrictions and many others

Mechanical design rules usually involve satisfying component cooling requirements, component mounting, vibration and shock protection, board dimensions, and physical mounting of the PCA in the system. Other layout rules, such as noncritical trace widths and spacings, via construction, hole-to-pad ratios, test-pad features, and part orientations and spacings, usually are standardized based on the manufacturing, assembly, and test processes to be used and reliability, maintainability, and cost constraints placed on the design.

3.4.5 Transition from Schematic to Layout

The primary input to the board layout process is the schematic diagram. Along with a parts list, it specifies what components are to be used and how they are to be interconnected. A schematic can be created either manually and provided to the layout function as a drawing or generated with a CAD tool. The latter approach is preferred because it avoids many of the potential errors caused by manual translation of the data required to produce a layout.

Many PCA design organizations that use automated board layout tools will capture manually prepared schematics into a CAD system environment before starting a layout. Once the captured schematic data have been verified, the self-checking features built into the design system will greatly reduce the chances for error as the layout is created.

Developing a layout in a completely nonautomated mode (manual schematic, manual layout) is usually accomplished by selecting or creating the appropriate dolls, as described earlier (Sec. 3.4.1), placing them on a scaled, gridded format, and drawing or taping the interconnections. As the interconnections are placed on the layout, the lines on the schematic that represent them are marked out. When all the connections are completed, all the lines should be marked. If the circuit is a simple one, drawn on a single-sheet schematic, probability of error is fairly low. If, however, the layout designer has to deal with a large, multisheet schematic, with many interconnections passing between sheets, the potential for creating a layout error increases significantly. If a multilayer board is being designed, the opportunity for making a mistake in the layout goes up astronomically. It is therefore imperative that a careful check of the finished layout be done by someone other than the original designer. The checker must verify that the dolls used correctly represent the required parts, that the layout rules were implemented properly, and that all interconnections were completed as described on the schematic.

Preparing a layout in a completely automated environment (schematic and layout) provides a designer with a number of controls and checks that are built into the system being used. They usually include, at a minimum, automatic verification that the correct parts are being used, the design rules are being adhered to, and the interconnections are being properly made. Most CAD systems perform these checks as the design is being created, and unless a designer specifically instructs the system to ignore an error it finds, the system will maintain a one-to-one relationship between the schematic and the layout and not allow any mistakes to be made.

3.4.6   Part Placement

Placement is one of the most critical steps in the board layout process. It affects how well the interconnections can be made; the manufacturability, testability, and maintainability of the PCA; the performance of the circuitry; and its response to environmental conditions. Part placement is considered to be an art by many designers, and "good" placements result from their prior experience and knowledge. Many PCA CAD systems include an autoplacement capability, but they usually apply simplistic rules and do not reflect real-world requirements.

There are very few standard rules and guidelines for placing parts on a board, since each design has a unique set of requirements that influence the layout. The strategy usually used by experienced designers is to manually locate critical components first to satisfy specified functional, proximity, and thermal requirements. This placement usually influences where the rest of the parts should go. At this point, autoplacement routines, if available, can be used to place these less critical parts. The designer may now also predefine the location of critical signal and power connections as part of the placement process.

If a CAD system is being used that incorporates an autorouter, a designer probably will want to perform a test route as a way of evaluating a proposed placement. The location and quantity of unrouted interconnections are a very good indicator of problems with the placement, and it is much easier to move parts around on the board before the final interconnections are defined. At minimum, if an autorouter is not available or will not be used, the designer usually can display point-to-point connections as a set of lines (sometimes called a *rats nest*) to highlight congested areas or longer than desired interconnections that may indicate potential problems. This is another instance where use of design automation tools provides a distinct advantage over the manual layout process, where it is difficult to predict what problems will arise due to how the parts are placed. It is also difficult and time-consuming to move manually placed parts after a large percentage of a board's interconnections have been established.

3.4.7 Interconnection Methodology

Part placement should be completed before the interconnection process is started; however, as routing of connections progresses, minor repositioning of parts may be desirable to enhance the design for performance or manufacturability improvement. A designer also should decide on how features associated with the interconnections are to be implemented and the strategy to be used for completing the hookup. It should be understood that as routing of the connections proceeds, these decisions may have to change to allow the design to be finished successfully. Some of these considerations are listed below, not in any order of importance, because their criticality is a function of the type of PCA being designed.

- Position and configuration of critical interconnections and circuit functions
- Power-distribution method and configuration
- Layer assignments for signals/power/ground, specific nets, trace direction (x, y), external/internal routing
- I/O connector, pin assignments, conductor routing
- Test point locations
- Conductor widths and spacings
- Component pin termination rules (unused pins, power/ground attachment)

- Routing rules (stubbing, length restrictions, proximity, crosstalk control)
- Keep-out area and edge-spacing requirements
- Methods for connecting to devices with high pin counts or closely spaced leads
- Shielding and circuit filter requirements
- Via implementation (placement, through-hole/blind/buried configurations)
- Thermal paths for heat dissipation
- Location of mechanical constraints (stiffeners)
- Manufacturability (conductor dimensions to compensate for board-fabrication processes, conductor balancing on layers, conductor density distribution, conductor corners, large conductor areas, land/hole ratios, lead/hole ratios, minimum annular ring, thermal relief pads on conductor planes, nonfunctional pads)
- Marking content, location, and methods
- Use of solder masks and conformal coating

In addition, the designer needs to determine if one or more quality conformance test coupons are needed to be associated with the circuit pattern and, if so, where they are to be located and what should they contain. IPC-2221/2222 provides a complete set of guidelines to aid a designer in configuring test coupons.

3.4.8 Layout Verification

Every physical layout should be checked, preferably by someone other than the designer, to verify that it is correct. Where and when in the design process checking is done and how it is accomplished depend mainly on what methods and tools were used to create the design.

Manually prepared layouts require checking procedures that are very different from those used for layouts that were developed with computer-based (CAE/CAD) tools. In either case, the design data should be inspected carefully to verify the correctness of the mechanical description, the relationship between the schematic and the interconnections on the board, the physical and circuit description of the parts, and the information provided in the manufacturing drawings. This subject is covered in much more detail in Chap. 6.

3.4.9 Documentation and Data

A set of drawings, phototools, and other data is normally required to fabricate, assemble, and test a PCA. The primary source of information for a data set is the completed and checked layout. Guidelines for producing this material can be found in IPC-D-325.

A drawing package for a PCA usually contains an assembly drawing, a parts list, a schematic, a board detail drawing (describing all mechanical features, board construction, and material requirements), and a master pattern drawing (a copy of the artwork masters). On some programs, the board detail and master pattern may be combined in a single drawing.

If a layout was created using a printed circuit CAD system or captured as computer-readable data by digitizing it, the required drawings usually can be produced by inputting the data into a computer-aided drafting program. Otherwise, the required drawing set must be created using manual drafting methods. In many cases, the bulk of the manual drafting work can be accomplished by making reproducible copies of the layout and artwork masters and pasting them onto standard drawing formats.

In addition to the drawings, a broad variety and structure of data may be required for manufacturing the bare circuit board and assembling and testing the PCA. This may range anywhere from simply providing a set of artwork films to creation of a set of data files that will be used to control automated fabrication, assembly, and test equipment. Most printed circuit CAD systems have the ability to create these types of data files in industry-standard formats and output them in a variety of media. A typical set might include

- A photoplot file to create artwork tools
- A file containing drill data
- Board routing/profiling data
- Data for automated part placement equipment
- Bare board test data
- Assembly test data

Many board fabrication and PCA assembly facilities can accept these data on a variety of media, including magnetic tape, floppy disk, CD-ROM, and by phone modem, either directly or through the Internet. A designer should determine what type of data are needed to build and test a PCA and the data formats that the manufacturer is capable of using. Figure 3.16 shows the interrelationships between the various drawings and the data produced to support PCA manufacturing requirements. The subject of drawing and manufacturing data is covered in more detail in Chap. 7.

3.5   Board Fabrication

Although a board may be constructed in an almost infinite number of ways, usually it can be assigned to one of five basic type categories: single- or double-sided without plated holes, double-sided with plated holes, multilayer, flex, or rigid-flex. Each type requires a unique fabrication process.

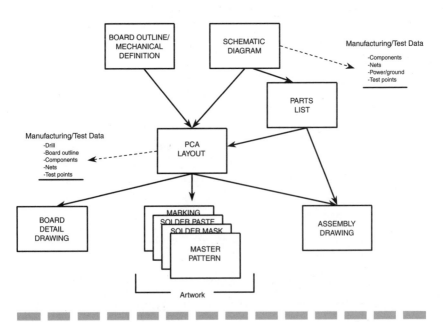

Figure 3.16
Manufacturing drawing and data relationships.

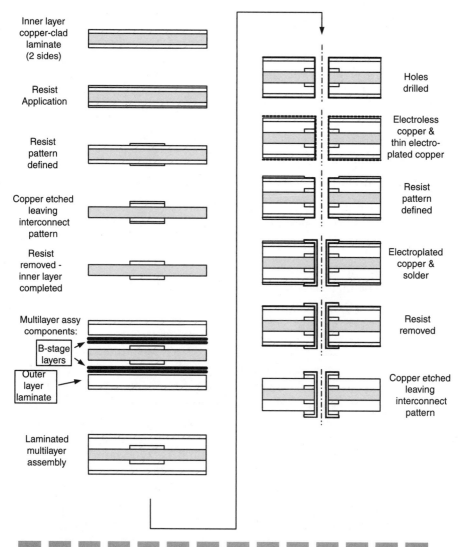

Figure 3.17
Multilayer board fabrication process using pattern plating.

The method selected for defining a circuit board's conductor pattern also determines the type of process to be applied. Pattern plating is the most common, followed by panel plating and additive processing. Figure 3.17 shows the basic steps normally used to fabricate double-sided and multilayer circuit boards, both using pattern plating with plated through-holes.

3.5.1   Process Selection

The exact process used for fabricating a circuit board is based mainly on its design features and associated requirements, including

- Construction and materials
- Physical dimensions
- Types of components used and how they are attached
- Circuit performance
- Component and interconnection density
- Interconnection placement and dimensional constraints
- Operating environment
- Quantity required
- The manufacturer's capabilities
- Cost and schedule constraints

Although details may vary from fabricator to fabricator, de facto industry-wide processes have been established based on the availability and capabilities of standard board materials, processing chemicals, and manufacturing equipment. Because of this, a designer generally can predict how a board will be made and therefore can ensure that the requirements associated with a design are compatible with the processing method.

If possible, a designer also should become familiar with the capabilities and constraints of the facility that may be used to make the board being designed and the details of its processes. If it is possible to tailor the layout to be compatible with a fabricator's capabilities and do so without having a negative effect on the quality of the design or limit its producibility by other suppliers, then the end result should be a higher-quality, lower-cost product.

3.5.2   Imaging

This step in the board fabrication process defines features on the surface of printed circuit board material using phototools generated for this purpose. The main purpose of imaging is to delineate conductor patterns, but it also can be used to apply markings, solder resist, solder paste, and other materials.

The process for defining circuit board conductors, as shown in Fig. 3.18, is relatively standard throughout industry. However, the type of phototools required and the type of resist material used will vary according to the construction of board being built and the configuration of the pattern defined by the layout.

Imaging is basically a photographic process, involving application of a photosensitive material to the surface to be imaged, exposing the material by shining a light through a phototool, and developing the resulting image. After removing unwanted material, what remains becomes a resist that covers the surface that is to remain unchanged during subsequent processing. If a surface is to be chemically etched away, then the resist covers the pattern that remains after etching. If a surface is to be overplated, then the resist covers the area that remains unplated. Conductor pattern imaging is one of the most critical steps in the board fabrication process; if done poorly, it can significantly affect the quality of the bare board and, ultimately, the performance of the PCA. Successful implementation of the imaging process not only depends on how well it is performed but also is directly related to the quality and accuracy of the phototools used to define the circuit pattern.

A resist pattern also can be applied directly to a surface using a screening process, but the screen itself is usually prepared photographically from phototools. The advantage of screening to a fabricator is that it is a less costly operation than using photoresist and requires a much lower capital investment. However, its ability to define fine-line patterns and hold close tolerances reliably is poorer. Screening is appropriate for applying markings (part designators, assembly numbers, etc.) on a circuit board, as well as for the application of solder resist, adhesive for chip part attachment, or solder paste. Again, if high-density, close-tolerance features are required, photo-imageable or machine-applied materials may be used.

3.5.3 Drilling

Holes must be provided in a circuit board for mounting and interconnecting leaded parts and as a path for making circuit

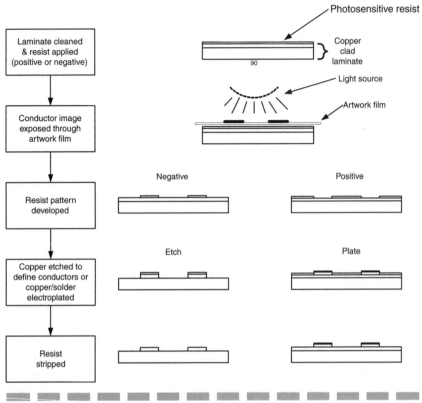

Figure 3.18
Conductor pattern imaging process.

connections between conductor layers. In addition, tooling holes and holes for attaching mechanical parts are usually needed. The locations and sizes of these holes are derived from the board layout and are specified on a board detail drawing.

A layout designer should consider and control a number of related factors when specifying hole dimensions and locations. Board construction and materials, the types of parts used, and the process capabilities of the board fabricator influence many of these factors. They include

■ Minimizing the number of different hole sizes used

■ Selecting hole sizes that can be made using standard diameter drills

■ Plated and unplated hole features

- Aspect ratio (board thickness to hole diameter)
- Hole diameter and positional tolerances
- Hole density (number of holes per square inch)
- Minimum allowable annular ring (external and internal layers)
- Whether hole breakout is permitted
- Minimum allowable spacing between adjacent holes
- Minimum allowable plated hole diameter

How hole features are defined has a major effect on the both the quality and cost of a circuit board. A complex, high-density PCA that uses parts with large pin counts or fine-pitch terminations usually requires a board design that incorporates small-diameter, tightly toleranced holes. Because of processing difficulty and reduced yields due to quality issues, the cost of this type of board will be much higher than one that uses standard dimensions and tolerances. Table 3.3 lists typical standard and high-precision values for some of the more important characteristics. IPC-2221 and IPC-2222 are good sources of information and general ground rules for defining circuit board hole characteristics.

It is also important that a designer understand the capabilities and limitations of the fabrication process and the potential sources of deviation from design requirements so that they can be accounted for in the design. Typical causes of error during drilling include mislocation, poor drill bit condition, excessive drill spindle runout, and drill wander or deflection. On multilayer boards, layer-to-layer misregistration is a typical source of quality problems, causing violation of annular ring requirements and hole/pad breakout that may result in open circuits or shorts between conductor paths. In addition, a marginal hole plating process may cause partial plating or no plating in small-diameter holes or holes with high aspect ratios. Elimination of nonfunctional inner layer pads on multilayer boards, if allowed, can help improve hole quality by decreasing the heat that is generated by a drill.

3.5.4 Chemical Processes

A variety of wet chemistry processes are used to manufacture circuit boards. These could include cleaning, photo-imaging, developing and stripping, etching (both metal and dielectric), plating

TABLE 3.3

Key Drilled Hole Features

Feature	Standard Value, in	Precision Value, in
Minimum unplated (drilled through-hole diameter	0.0135	0.006
Minimum unplated (drilled) blind/buried via diameter	0.006	0.003
Unplated hole diameter tolerance	±0.002	±0.001
Hole location tolerance (radial true position)	0.004	0.002
Plating thickness in hole (total)	0.006	0.004
Aspect ratio of plated through-hole	5:1–8:1	9:1–12:1
Minimum spacing between adjacent holes	Hole diameter	$\frac{3}{4}$ hole diameter
Minimum distance between edge of drilled hole and circuitry	0.010 external 0.013 internal	0.005 external 0.007 internal
Minimum pad diameter > drilled hole	0.020 external 0.015 internal	0.015 external 0.010 internal
Minimum allowable annular ring	0.005 external 0.003 internal	0.004 external 0.002 internal
Plated hole to lead diameter relationship	≤0.028 over min lead diameter ≥0.012 over max lead diameter	≤0.020 over min lead diameter ≥0.005 over max lead diameter

(electroless, electroplating, additive), fluxing, solder leveling or fusing, and application of solder resist. A designer should be aware of the general properties of the chemicals that the board materials will be exposed to during processing and the conditions of exposure (mainly thermal and mechanical stresses that could cause board warpage, delamination, excessive moisture absorption, and corrosion). The board material and construction specified in the design must be capable of withstanding this exposure without degrading any of its physical or circuit characteristics.

A designer also should be aware of the limitations of some of these wet processes and how they may constrain layout features.

The ability to accurately define fine-line conductor widths and spaces is directly affected and limited by the resist process used (material and method of application) and the etching and plating processes. As mentioned earlier, plating may not meet quality requirements in small-diameter, high-aspect-ratio holes, and their use in a design also may be limited or not allowed. Careful selection of board fabricators is extremely important, especially if a board design specifies fine-line conductor widths and spaces in the 3- to 6-mil range and plated hole diameters of 12 mils or less.

3.5.5 Lamination

Multilayer boards are constructed by laminating or gluing together individual layers of laminate material on which interconnection patterns have been predefined. Adhesion between the layers is provided by interleaving them with partially cured sheets called *B-stage* or *prepreg material,* which usually consists of a combination of partially cured resin and a reinforcement material. During the lamination process, heat and pressure are applied to the sandwich that bond the circuit layers together by remelting the B-stage material, and then the board is allowed to fully cure. Figure 3.19 shows an exploded cross section of a typical multilayer board stackup and the finished board dimension after lamination.

Layout designers should be aware of the flaws resulting from lamination that may affect the quality and performance of a finished board. Some of these potential process-related defects are blisters, voids, delamination, measling, trapped debris, misregistration, incomplete fill, and thickness deviations (between layers, finished board, variability across the board)

It is possible to configure a layout to avoid some of these problems or at least minimize their impact on board quality. Actions by a designer that could enhance the manufacturability of a multilayer board during lamination include

- Avoid large conductor areas. If these are mandated by circuit performance requirements, then they should be broken up by adding a dot or slot pattern to enhance adhesion.

- Avoid areas of high conductor density. Spread traces wherever possible.

- Avoid large fill areas that may cause localized adhesive depletion.

- Voids usually form at the edges of a panel. A board outline should be no closer than a half inch from an edge.

- Avert board warpage by locating layers of laminate and B-stage material so that they are symmetrical around the cross-sectional centerline of a stackup. Also orient the direction of conductors at right angles to each other on alternating layers.

- Select laminate and prepreg material dimensions and tolerances that will result in the correct layer-to-layer and overall board thickness. Impedance, crosstalk, and voltage breakdown parameters are directly related to the thickness of dielectric material between conductor layers (e.g., the impedance value tolerance equals the dielectric thickness tolerance).

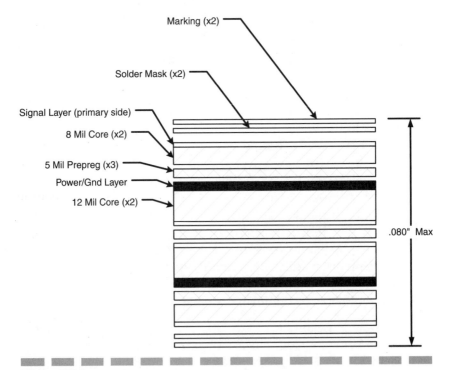

Figure 3.19
Typical multilayer board stackup and allowable finished dimension.

■ Specify an overall board thickness tolerance of ±20 percent. A tighter tolerance will require special processing, resulting in a more costly product.

3.5.6 Marking

Markings are placed on a circuit board for two primary purposes: to identify the board (both detail part and finished assembly) and as indicators for placement and orientation of component parts (reference designators, polarity marks, etc.). Identification nomenclature should be structured to provide traceability to the version of the documentation used to manufacture the bare board and final assembly. This can be done by appending a dash number or revision letter to a base name or part number. Component reference designator and orientation markings are derived from pertinent information in the schematic and on the layout.

Markings must be legible and be able to withstand all manufacturing processes and environmental conditions to which they are exposed after they are applied to a board. Reference designators and orientation marks should be visible after parts are installed to aid in performing test and repair activities.

A board may be marked by etching characters into or from a conductive surface, by applying ink to a surface, or with a combination of both processes. Ink may be screened, stenciled, or stamped on. For precise placement of markings on complex, high-density assemblies, screening or stenciling is preferred. Artwork films are required to produce the screening or stenciling tools. These may be produced directly from a manual layout and photographically reduced or photoplotted as 1:1 artwork. Marking ink should contrast with the color of the laminate and be nonconductive, noncorrosive, and resistant to environmental extremes. Epoxy-based inks are commonly used.

Care should be taken by the designer when placing markings on a board to avoid compromising its performance. Potential problems include

■ Ink bleeding into plated holes causing solder voids

■ Ink partially covering circuit pads causing unacceptable solder joints

- Etched nomenclature that violates minimum spacing requirements

- Etched nomenclature that separates from the laminate causing a short between conductors

- Markings that cross over circuit features resulting in illegible characters

- Ink bleeding over the edge of solder mask openings

- Illegible markings due to specification of character size and weight beyond processing capabilities

3.5.7 Solder Mask

The main reason for including a solder mask on a circuit board is to control the application and flow of solder employed during component assembly. If it is a permanent solder mask, it also can provide environmental protection or insulation between closely spaced conductors and prevent electromigration of conductive material. Proliferation of circuit board designs incorporating dense circuitry and surface-mounted components with fine-pitch terminations has greatly expanded the use of solder masks as a way of mitigating the associated processing challenges. Figure 3.20 shows a typical solder-masked circuit board.

The decision to use a solder mask is mainly cost-related and involves trading off additional material and processing costs against the potential reduction in yield and increased inspection and rework that may be required if it is not used.

Material can be applied as either a mask that becomes a permanent part of a circuit board or as temporary resist to be removed prior to completion of the assembly. A permanent solder mask is specified as an element of the design, and its material and configuration requirements are defined in the manufacturing documentation and data provided to a board fabricator. Temporary masking is applied and removed as part of the manufacturing or assembly process and is of no concern to a designer as long as it has no effect on the performance of the finished product.

A broad variety of solder mask materials and associated application methods are available and widely used throughout industry. The criteria for choosing an appropriate material depend on

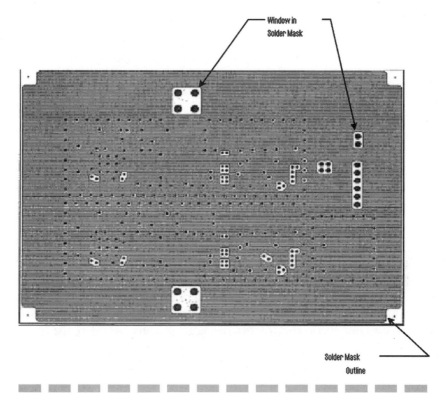

Figure 3.20
Typical solder-masked circuit board.

the design features of a PCA and its performance requirements. Key selection criteria include

- Compatibility of electrical and physical properties with those of the board laminate and, if used, conformal coating material

- Ability to withstand operating conditions without degradation

- Compatibility with fabrication, assembly, and rework/repair processes and materials (heat, solvents, fluxes, cleaning materials, abrasion, soldering)

- Ability to provide the appropriate material thickness, feature dimension resolution, and placement accuracy

- Ability to tent over holes to prevent solder fill, if required

Material thickness and feature resolution are critical solder mask characteristics and are interrelated (thinner material allows finer feature resolution). High-density circuit board layouts require that very tight control be exercised over the position and dimensional accuracy of solder mask features. Mask openings should be dimensioned so that material uniformly surrounds a circuit pad but does not cover any part of it under worst-case tolerance conditions. The mask must prevent solder migration but not inhibit formation of an acceptable solder joint.

Control of mask thickness and feature location is also associated with proper application of solder paste for attachment of surface-mounted components. Solder mask openings and screen or stencil openings must be closely aligned to ensure that paste is properly deposited on each component pad. If the height of the mask is much greater than the pad thickness, accurate paste deposition becomes more difficult.

There are a number of other process-related issues that a layout designer should consider when specifying a solder mask. Covering thick conductors may cause voids between the mask and the laminate, resulting in entrapment of corrosive material. Also, the resulting uneven surface may affect the quality of markings in these areas.

Another potential problem involves melting of solder on conductive areas covered by the solder mask. This may be caused by heat applied to a board during a mass component soldering operation and can cause significant damage to the solder mask material. To prevent this, a designer may have to specify that these areas not be solder-coated, allowing a board fabricator to use a solder mask over bare copper (SMOBC) process.

The methodology and criteria that can be used for selecting the type of solder mask to use in a design are described in IPC-SM-840. Mask material is basically a polymer that can be applied as a liquid or a solid film and cured by the application of heat or ultraviolet energy. Its features (openings) are defined by applying the material through a screen or stencil or by a photographic process similar to that used for defining conductors. Artwork data must be provided either to produce the screen or stencil image or for preparing phototools needed to expose the mask image on the board. Table 3.4 lists various types of solder masks, their thickness and feature resolution capabilities, and the types of processes used to apply them.

TABLE 3.4

Features of Commonly Used Solder Mask Materials

Type	Thickness Over Laminate, mils	Feature Resolution, mils	Application Process
Screenable	3.5–5	10	Screen or stencil, heat or UV cure
Dry film	2.3–4	6	Laminate, expose, develop
Liquid photo-imageable	1–1.7	<6	Curtain or screen, expose, develop

3.5.8 Inspection and Electrical Testing

Visual inspection and electrical testing are the two primary methods used to verify the quality of a fabricated board. Visual inspection is done to ensure that all a board's physical features meet specified requirements. A deviation from requirements is usually categorized in one of two ways. It can be identified as a defect that must be repaired or, if it cannot be fixed satisfactorily, causes the board to be scrapped. The other type of deviation is one that may not violate bare board specification requirements but may result in future degradation of PCA performance. Table 3.5 lists potential physical board defects in each category.

Physical defects may be found using one or a combination of methods, including

- Manually scanning the external surfaces of a board and the individual inner layers of multilayer boards
- Using optical aids
- Cross sectioning a quality conformance test coupon and examining the specimens under a microscope
- X-raying a board to view defects on inner layers

Unfortunately, as board complexity increases, the ability of the human eye to consistently resolve and identify surface defects goes down radically. A solution to this problem is to use automated optical inspection (AOI) equipment. Accept/reject

TABLE 3.5

Physical Deviations from Requirements

Defects that *must* be repaired	Defects that could cause future performance degradation
Conductor shorts/opens	Surface scratches (conductor, dielectric)
Conductor/pad dimensional or space violations (incl. excessive undercut, neckdowns, "bites")	Surface dents, pits, roughness
Registration/annular ring violation	Measling/crazing
Surface voids/pinholes (conductors, lands, plated holes)	Plated hole roughness
Plated hole voids (>10% of plated area, finished diameter violation, plating, thickness, pad separation)	Solder mask defects
Conductive whiskers/slivers	
Foreign particle inclusions (>0.1 mil)	
Solder mask violations (pad coverage)	
Surface plating thickness violation	
Delamination (interlayer, conductor)	
Weave/fiber exposure	
Bow or twist violation	
Solderability (adhesion, wetting)	
Illegible marking	

parameters can be programmed into an AOI system, allowing it to identify rejectable defects such as conductor shorts or opens and marginal ones such as conductor neckdowns or poor surface conditions. A typical AOI system is shown in Figure 3.21.

Electrical testing is the best method for identifying conductor defects on a board. Using a preprogrammed tester that makes contact with conductors or pads on a board's surface, the continuity of each circuit net can be checked, and its isolation from all other nets can be verified. The information required to

Figure 3.21
An example of a circuit board AOI system. (*Used by permission of Barco ETS.*)

program such a tester can be derived directly from data files produced by the CAD system used to lay out the board or generated with a fabricator's computer-aided manufacturing (CAM) system. The data required to make a board test fixture also can be produced in this way. Figure 3.22 shows a typical bare board tester.

3.6 Assembly

The technology used for mounting and attaching components to a circuit board determines the basic process that will be employed to produce a finished assembly. A PCA may use through-hole technology, surface-mount technology (SMT), or a mixture of both.

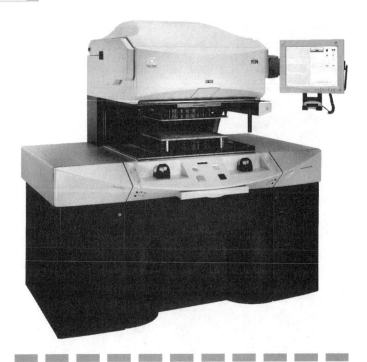

Figure 3.22
Bare circuit board tester. (*Used by permission of Everett Charles Technologies.*)

Through-hole assemblies are the easiest and also may be the least costly to produce. This technology is usually applied in a design where there are no size constraints or where a circuit requires a majority of parts that are only available in a through-hole configuration (such as transformers, large capacitors, inductors, connectors, etc.). Through-hole boards can be assembled manually or with automated equipment.

Assemblies that use surface-mount technology (SMT) or mixed technology are much more difficult to build, requiring complex processes that are implemented with costly, automated equipment. While through-hole assemblies usually have parts mounted to one side of a board, SMT or mixed assemblies may have parts on both sides. Also, the physical size and configuration of surface-mounted packages and their terminations require that the processes used to attach and connect them to a circuit board be very tightly controlled.

When selecting the types of parts used on a PCA and determining where they are placed on the circuit board, a designer should consider the following:

■ Availability and cost of components that satisfy circuit performance, quality, and reliability requirements

■ Assembly, test, and handling requirements

■ Component density and orientation requirements

■ Compatibility with automated processes

■ Quantity of PCAs to be made

■ Cost constraints

■ Ability to survive exposure to assembly and test processes and materials without jeopardizing the quality and reliability of the finished product

■ Repair and maintenance requirements

■ Use of industry-standard assembly processes

3.6.1 Assembly of Through-Hole Components

Through-hole parts are available as discrete components with axial or radial leads, components with in-line leads, large multipin parts, and nonuniformly shaped parts. Before a layout is started for this type of assembly, it should be determined if parts will be mounted on the circuit board using manual or mainly automated processes. This decision depends on many considerations but usually is made based on which method is more cost-effective, assuming that technical requirements can be satisfied with either approach.

There are many more layout constraints associated with designing a PCA for automated assembly than are required when using manual assembly methods. If it is determined that machine insertion is to be used, a designer should ensure that the layout incorporates features needed to make the PCA producible in that environment. These include

■ Selecting parts that are physically compatible with pick-and-place equipment

■ Minimizing the number of different part shapes used

- Orienting parts parallel or perpendicular to each other and in the direction of a board datum
- Minimizing part orientations (direction and polarity)
- Allowing sufficient space between parts for tool footprints (on both sides of the board)
- Minimizing the number of manually inserted parts and hardware
- Mounting all parts on one side of the board
- Specifying uniform lead forming and clinching requirements
- Providing adequate hole diameters and tolerances to ensure reliable lead insertion

3.6.2   Assembly of Surface-Mounted Parts

To satisfy the call for more complex, highly miniaturized electronic products, PCA designers are increasingly turning to the use of surface-mount technology. Most assemblies now being produced apply this approach either by itself or combined with through-hole technology. Use of surface-mounted parts also offers significant cost advantages over though-hole parts in the following areas:

- Elimination of lead preparation activities
- Ease of shipping and handling
- Suitability for high-rate automated assembly
- Compatibility with mass soldering processes

Over the last several years, availability of electronic components in surface-mount packages has improved to the point where almost every type of standard part can be obtained in this configuration. Circuit performance of surface-mounted parts is usually equal to its through-hole package equivalent, and for high-speed circuitry, performance is better due to the reduction of the length of conductor paths between the circuit element and the outside world.

There are a number of critical technical issues that a designer should consider when using surface-mount technology. Because of the smaller size of the devices, power densities are increased, and more attention must be given to heat removal. Also, there could be a mismatch of thermal coefficients of expansion (TCE) between the device package material and the circuit board material. This may cause terminal solder joint failures due to mechanical stresses on the joints during temperature cycling, especially if unleaded parts are being used.

Other considerations include

- *Package material.* Plastic packages will provide a better CTE match with the circuit board than ceramic packages, but they are more susceptible to performance degradation when exposed to high temperature and humidity.

- *Terminations.* Leads on SMT devices are usually fragile, can be bent or skewed easily, and may require special handling and preassembly inspection. Also, termination materials and plating may significantly affect the quality of associated solder joints (wetability, formation of nonconductive intermetallics, etc.). When choosing between leadless versus leaded packages, other selection tradeoff issues may involve cost, space required, and access for test and solder joint inspection.

- *Fine-pitch SMT parts.* Fabrication and assembly capabilities and costs are factors of concern. Interconnection density (fine-line pad and conductor widths and spacings) affects board yields, and higher placement accuracy requires use of more sophisticated assembly equipment. Also, access for circuit testing and rework and repair becomes an important issue.

- *Land patterns.* Each type of SMT device requires a unique land pattern for interconnection to a circuit board. The producibility of a PCA using SMT is very dependent on the design of the land patterns. Guidelines for developing land patterns are contained in IPC-SM-782.

- *Autoplacement.* Most of the guidelines described earlier (Sec. 3.6.1) for through-hole technology are also applicable to SMT devices. In addition, fiducial markings should be added on the board to provide an optical reference for accurate placement of parts with large numbers of terminations or fine-pitch spacings.

- *Part attachment.* For most SMT parts, their soldered terminations are sufficient to keep them attached to the circuit board. For designs that have parts mounted on both sides of a board, the devices on the bottom (secondary) side must be bonded to the board with an adhesive to hold them in place during the soldering operation. There are also instances where SMT devices are bonded to a board to enhance heat removal by providing direct thermally conductive paths.

SMT components are available in a large variety of form factors, package materials, and termination configurations, including a number of less widely used package configurations such as chip-on-board (COB), tape-automated-bonded (TAB), and flip-chip devices. When considering the use of these packages, the designer should be aware of their cost and availability and the need for special equipment and processes to mount them.

3.6.3 Soldering

The types of parts used, how they are mounted, and the predicted use and operational environment of a PCA will affect the soldering process and the soldering materials specified for a design. Table 3.6 summarizes the soldering methods typically used for different board technologies. Although the details of how they are implemented will vary, the two basic techniques used throughout industry for mass or machine-made interconnections are wave soldering and reflow soldering.

The former involves passing a circuit board with components assembled on it across a molten wave that adds solder to make the interconnections as the board leaves the wave. Reflow soldering requires that solder already present on the assembled board be heated to its melting point, which when cooled forms the interconnections. As shown in the table, some types of technologies require the use of both methods on the same assembly. Another common soldering technique involves dipping an entire assembly into a molten solder bath to make the interconnections, but the process is difficult to control, and its use is not widespread for production.

It is important that the designer know what type of soldering process will be used for a specific PCA so that the layout can be configured to accommodate the tools and fixtures required for

TABLE 3.6

Soldering Methods Typically Used for Different Types of Assemblies

Assembly Type	Soldering Process
Through-hole components mounted on a single side	Wave solder
Surface-mounted components on a single side	Reflow solder
Surface-mounted components on both sides	Reflow solder or reflow and wave solder
Mixed technology, through-hole and surface-mounted components, on a single side	Reflow and wave solder
Mixed technology, through-hole components on one side, surface-mounted components on both sides	Reflow and wave solder

that process. Features also should be incorporated in the design for compatibility with the process and to ensure that the PCA's performance is not degraded by the soldering operation. Considerations should include

■ Selection of solder materials compatible with the board conductor and part termination surface materials and finishes

■ Specification of flux materials whose residue after soldering is minimally corrosive and can be removed completely using environmentally acceptable cleaning processes

■ Selection of electronic components and mechanical parts that will not be degraded by the materials and temperatures associated with the soldering process

■ Specification of board conductor and component termination surface conditions for solderability (cleanliness, pretinning, gold plate removal, etc.)

■ Definition/description of acceptable solder joint quality and methods to be used to verify that level of quality

■ Use of conductor land patterns and adequate lead-to-hole diameter ratios that will ensure creation of acceptable solder joints

- Mounting of electronic components to prevent termination movement during or immediately after formation of interconnections that may cause creation of "disturbed" solder joints having a high potential for subsequent failure

- Minimizing the possibility of solder joint failure due to fatigue caused through stresses generated by temperature cycling during PCA operation due to large differences in the coefficients of thermal expansion (CTE) of the circuit board and the soldered parts

- Mounting of parts to allow cleaning and removal of flux residue

- Use of appropriate solder masking materials and geometries, especially for fine-line, fine-pitch circuitry or if via tenting is required

- Selection and application of solder paste materials and provision of artwork for required process tooling

- Minimization of hand soldering and manual solder joint touch-up operations

3.6.4 Markings

A variety of markings and legends must be included and defined as part of the PCA design process. These include component reference designators, board and assembly part numbers, serial numbers, revision levels, orientation and polarization indicators, alignment and registration marks, and many others. Methods used for producing them vary depending on what type of marking is required.

Fixed-format information is usually included in the master artwork. This includes board and assembly part numbers, circuit layer numbers, reference designators, orientation/polarity/alignment markings, industry standards (such as UL), and electrostatic discharge (ESD) warning and sensitivity labels.

Variable information is usually added at various points in the manufacturing process using permanent ink, labels, laser scribing, or other methods. This type of information typically includes serial numbers, date codes, lot codes, and manufacturer's designation.

The material used for marking should be a nonconductive, nonnutrient, high-contrast substance (usually an epoxy-based ink). It should be chemically compatible with and adhere to the other materials it will contact, including the circuit board and solder mask or conformal coating, if used. Stick-on labels are also sometimes used. These markings should be able to withstand the materials and processes used during manufacturing, test, rework and repair, and field maintenance and, ultimately, survive the PCA operational environment without noticeable degradation.

If quality conformance test coupons are produced as part of the board fabrication process, they also must be marked. The marked data should include the board part number and revision, the coupon identifier, the lot and date code for the board, and the manufacturer's identification.

The ability to revise markings should be provided. If a circuit board or assembly is modified, then the appropriate marking must be changed in order to properly identify the new configuration. At minimum, a change of this type usually will affect the board or assembly part number and the revision level.

Markings should be located for maximum readability and be visible after the assembly has been completed. Minimum preferable dimensions for marking are 0.060 in high with a line width of 0.012 in. Space required for markings should be considered during component placement. Avoid locating markings under parts, on conductive surfaces, or on surfaces that will melt or be covered with an opaque material during subsequent assembly operations. Inked characters or labels should not cross more than one conductor. Care should be taken to prevent entrapment of moisture under stick-on labels. Markings should not be allowed to touch surfaces that are to be soldered to prevent nonconductive material from leaching into the joint during the soldering operation.

3.6.5 Assembly of Mechanical Parts

Almost without exception, PCAs will use mechanical parts and hardware. Typical mechanical parts incorporated in a design can include

- Supports and stiffeners
- Component tie-downs such as clips, clamps, brackets, and straps

- Heat sinks and thermal interfaces to external structures
- Board extractors
- Terminals, eyelets, and rivets
- Spacers and insulators

Supports and stiffeners usually are included in a design to provide rigidity to an assembly and prevent damage from flexing due to shock and vibration during operation or handling while maintenance or repair is performed. Stiffeners are also added to large, unsupported boards to reduce bowing, especially if edge connectors are used. When specifying metal stiffeners, care should be taken to select material and surface finish compatible with the rest of the design and suitable for its eventual operating environment. In addition, care must be taken to ensure electrical isolation from the board circuitry. When establishing clearance between conductive parts and board circuitry, care should be taken to consider all possible mechanical tolerances associated with the part, the board, and the mounting hardware.

Installations of component mounting hardware (clips, clamps, etc.) should be configured to prevent movement (translation or rotation) that could put damaging stress on the part or its associated solder joints. In addition, movement of a conductive mechanical part could cause a short circuit.

When specifying the installation of mechanical parts and hardware, the designer should be aware of some basic issues and requirements that are unique to these types of components. How the attachment of permanent hardware is specified is particularly important. The forces that are required for setting of flanges on terminals, eyelets, and rivets should be controlled so that there is no resulting damage to the circuit board conductors or dielectric material due to cracking or crushing. If a terminal, eyelet, or rivet does not form part of the electric circuit on a board (not soldered to a conductor), its flange may be rolled. If it is part of the circuitry, the flange should be flared to ensure that a soldered connection is properly made. Figure 3.23 shows examples of both types of installations. When specifying press-in terminals that make circuit connection by being inserted into a plated hole, care must be taken to prevent damage to the circuit by selecting the correct hole size and plating thickness. The terminal vendor should provide this information.

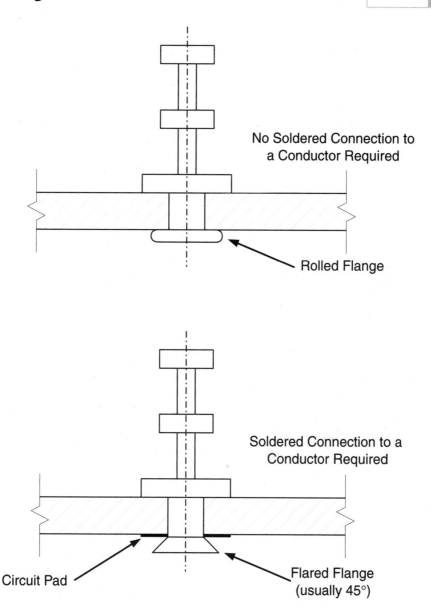

No Soldered Connection to
a Conductor Required

Rolled Flange

Soldered Connection to a
Conductor Required

Circuit Pad

Flared Flange
(usually 45°)

Figure 3.23
Examples of solder terminal installations in a circuit board.

The forces involved in the attachment of fasteners to a circuit board also should be controlled carefully. A maximum allowable tightening torque should be specified for each type and size of fastener to reduce the risk of crushing the board and affecting its subsequent performance. Fastening mechanical parts to a board using lugs, twist tabs, or ears should be avoided. It is difficult to control the forces involved in mechanically setting these features. Making them too tight will cause damage to the board, and making them too loose eventually will result in the loss of functionality of the attached part.

Mechanical stresses on a board also should be considered when designing provisions for board extractors. Plug-in PCAs incorporating connectors with large pin counts require large extraction forces for removal. The designer should ensure that the areas of the circuit board used for external extractor attachment or built-in extractors are sufficiently reinforced to prevent damage to the board. Also, adequate clearance for attachment and movement of the extractors should be provided in the layout.

Mechanical assembly of a heat sink on a circuit board is particularly critical, since it frequently requires establishment of a good thermally conductive path between a part and the board while precluding an electrically conductive path. Provision of an interface material between the heat sink and the board, such as thermal grease or thermally conductive adhesive, may be required to provide the heat-transfer path required for effective part cooling. Selection of the proper materials for compatibility with surrounding surfaces and manufacturing and assembly processes is as important as their thermal and electrical properties.

There are also some general issues to be considered when selecting mechanical parts and placing them in a layout. These include potential interference with other parts on a board and with adjacent assemblies, at what point in the assembly process they should be mounted on a board, and if manual or automated installation procedures are to be used.

3.6.6 Conformal Coating

Conformal coating is primarily used on a PCA to ensure its reliability by protecting circuitry from potentially damaging operating environments. It is especially applicable for complex, high-density designs. Properly selected and applied conformal coating will prevent degradation of performance due to the effects of humidity, dust and dirt, vibration and shock, corrosive chemicals, abrasion, fungi, and loose conductive materials. A secondary function performed by conformal coating is provision of physical support of components on a board, which relieves some of the mechanical stresses on their solder joints.

Key material properties that should be evaluated when selecting a coating material include

- Operating temperature range
- Solvent and chemical resistance
- Abrasion resistance
- Adhesive characteristics
- Flammability
- Dielectric breakdown voltage and bulk resistance
- Moisture resistance
- Fungus resistance
- Aging characteristics and long-term chemical stability
- Opacity
- Outgassing characteristics
- Method of application
- Repairability

It is also very important that the coating material specified be mutually compatible with the surfaces and materials it contacts, considering adhesion, chemical interaction, and matching of coefficients of thermal expansion. Some of the information needed to make the determination may be difficult to obtain, but the vendors of the involved materials should be able to provide it, and this evaluation should be done to prevent potential performance degradation or failure. Some materials require special surface pretreatment to ensure adhesion. Some may contain solvents or chemicals that will affect the integrity of markings and labels. Still others may involve use of application and curing processes that will adversely affect existing materials.

Conformal coating material should retain its physical and electrical characteristics at the extreme limits of the PCA's expected operating temperature range. At the low end it should be flexible enough to bend without cracking, and at the high end its dielectric properties should not be degraded. Its adhesive and cohesive characteristics should not be outside the manufacturer's specification limits throughout the expected range of temperatures to which the material will be exposed.

Another critical concern is repairability of conformally coated assemblies. Some materials are quite easily removed and replaced during PCA repair or modification, whereas some are very difficult to remove. Conformal coating usually can be removed mechanically (with cutting, peeling, abrading), thermally (with local heating), or chemically (with solvents). The primary requirements for the process and materials used for localized removal of conformal coating are that the surrounding components and circuit board not be damaged during removal and that the repaired area be able to be returned to its original condition.

Specific design requirements that are unique to the use of conformal coating should be observed and applied during PCA layout. These include

- Close control of material thickness to prevent damage to components and solder joints due to stresses generated during temperature changes by differences in CTE
- Complete coverage of surfaces to be coated with no voids or inclusions
- Provision of flexible buffer material as an interface between the coating and brittle component packages (glass or ceramic)
- Preventing space between flat components and the circuit board to be filled by coating material
- Specifying surfaces not to be coated (test probe areas, connectors, adjustable parts, ejectors, screw threads, card edge surfaces used to conduct heat from the assembly, heat-sink fins)

3.7 Test and Inspection

Testing and inspection are performed after a circuit board has been assembled to verify that it will meet its circuit and functional performance requirements and that it satisfies the quality requirements specified for the design. Circuit testing is done to confirm that all parts work properly and that they are interconnected correctly. Functional testing proves that the PCA will operate as required when it is installed into its intended system.

Additional testing is sometimes performed to validate the reliability of a design. This may be accomplished by exposing an assembly to a series of tests that simulate a combination of environmental conditions that it may be exposed to during its operational lifetime and observing its ability to function properly during and after this exposure. An assembly also may be operated in an overstressed condition in order to identify potential weaknesses in the design that can be corrected before the product is put into operation.

Inspection may be done at various points during the assembly process. Its purpose is to establish that the product meets all the quality requirements specified in the documentation that was used to build the PCA. A completed assembly is usually inspected visually, and emphasis is placed on confirming the proper formation of solder joints, component mounting, mechanical part installation, elimination of extraneous conductive materials, legibility of markings, conformance to dimensional requirements, and general cleanliness.

3.7.1   In-Circuit Testing

This type of test is used to confirm that the correct electronic parts have been assembled on a board, that their key circuit parameters are within acceptable limits, and that they are properly interconnected. In-circuit testing can be performed on both analog and digital devices, and the more sophisticated testers can perform limited functional tests on individual parts and groups of parts.

Electrical contact with components and circuitry on a board is usually made with spring-loaded pins that are mounted in a fixture and wired to electronic test equipment. The fixture is custom made with the pins precisely located to contact each component termination and interconnection net. To minimize the cost of the fixture, the board should be laid out so that all contacts can be made one side of the assembly. Test point contact areas should be located so that they are accessible from above the board. They also should be evenly distributed across the surface of the board to avoid the possibility of damage to the circuitry due to concentrated forces created by the spring-loaded pins.

Prior to starting a layout, the designer should identify requirements and rules for in-circuit testing so that accommodation can be made for placement and interconnection of test point contact areas. These include such considerations as probing of unused pins; measurement of net, power, and ground characteristics; test capabilities built into parts; functional testing to be performed; and others.

3.7.2 Functional Testing

The main objective of this type of test is to determine if a PCA will perform its required circuit functions when it is installed in its intended next-higher-level assembly. This usually involves connecting the PCA to a tester that simulates the operation of the final product.

The interconnection is made through the same interface that would attach the PCA electrically to its next assembly. If variable components with adjustable values are included in a design, this is usually when the final values are set. It is important that the designer be aware of the type of fixturing that will be used for functional testing and make sure that the layout will be compatible with it.

3.7.3 Visual, Dimensional, and Mechanical Inspection

As PCAs have been become more complex and circuit densities have increased, visual inspection has become a very important factor in ensuring that the assembly conforms to the requirements and specifications imposed on the finished product. Electrical testing is effective in identifying problems associated with electronic components, such as bad parts, wrong parts, missing parts, or incorrectly oriented parts. However, a significant percentage of assembly faults are associated with mechanical defects that usually are not identified by circuit testing but eventually can cause circuit failure during operation of a PCA. These include

- Marginal solder joints (voids, insufficient solder, "cold" joints, dewetted surfaces)
- Components misaligned with circuit pads
- Solder balls or other loose conductive materials
- Slivers or whiskers of conductive material that eventually may break away and cause shorts
- Partial conductive paths that eventually may close due to electrochemical migration
- Existence of potentially corrosive residue from assembly processes (such as solder flux) that may affect long-term circuit performance

The opportunity for these conditions to exist is further increased by the widespread use of components with high pin counts such as pin-grid or ball-grid arrays, where most of the solder joints are not easily viewed.

Visual inspection also should locate and identify "cosmetic" defects. Scratches, gouged or cracked areas, abrasions, rough or ragged edges, missing material, measling, cuts or nicks, charred or burned areas, and similar anomalies may be latent defects that could cause performance degradation or failure later on in a PCA's life cycle. Workmanship specifications and associated criteria imposed on a PCA are usually prescribed on the assembly drawing, and compliance with these requirements should be certified by visual inspection.

Conformance with dimensional requirements should be confirmed by visual inspection. Various assembly processes, if not properly controlled, may affect the physical characteristics of a circuit board, causing it to exceed its specified warp, bow, or twist limits and affecting the PCA's installation into the end product. An assembly also should be checked visually to ensure that electronic and mechanical parts have been mounted in accordance with the manufacturing drawing and that they do not violate any height or keep-out area restrictions. Proper spacing between conductive mechanical parts and hardware and circuitry should be verified, and areas surrounding torqued or formed fasteners should be examined to ensure that their installation did not damage the circuit board material.

If markings, adhesives, or conformal coatings are used on an assembly, they should be inspected visually for proper application. Marking and adhesive materials applied before soldering should not touch or overlap any surfaces to be soldered. Conformal coating should be measured to confirm that it does not exceed the maximum allowable thickness and checked to confirm that it forms a continuous film over all required surfaces and that it is not covering any area that it should not cover.

3.7.4   Environmental and Reliability Testing

Environmental and reliability testing is intended to predict the performance of a PCA design by exposing a sample number of assemblies to a combination of conditions in which the design may be required to operate. A design may be tested while installed in the final product or by itself. The tests are usually configured to expose a PCA to a combination of the worst conditions it will be expected to encounter during its life cycle, including shipping, storage, and operation.

Most environmental conditions can be created in a test chamber, and the chamber settings can be cycled many times to simulate product aging. A PCA can be tested without power applied to its circuitry and powered up afterwards to identify permanent effects of the environment or tested while it is operating so that its performance can be monitored under varying environmental conditions. Test conditions can include extremes of temperature, altitude, shock and vibration, humidity, rain, salt spray, and others.

Testing also can be used to predict a design's reliability during its expected performance lifetime. By continuously operating a PCA while overstressing it environmentally, potential failure mechanisms inherent in the design can be accelerated. Design weaknesses may be identified in this manner and corrected before full-scale production is started.

The designer should know what environmental conditions a PCA will be exposed to and make sure that the assembly will be sufficiently robust to survive them without appreciable degradation in performance. Environmental extremes and how they may be combined over the life of a product should be identified, carefully analyzed, and accommodated in the layout so that the PCA is neither overdesigned nor vulnerable to premature failure.

Circuit Board Layout

Over the years, successful circuit board layout activities have been guided by the so-called three-legged stool philosophy, with each leg representing full optimization of a printed circuit assembly's (PCA's): cost, schedule, and performance requirements. These three factors should be given equal consideration to ensure overall design integrity. Unfortunately, many companies place the greatest weight on "time to market" constraints and, by doing so, sacrifice one or two of the three legs. The consequence is that during the layout effort, primary emphasis may be placed on meeting a restrictive schedule, resulting in possible degradation of the cost-effectiveness or performance of the design. Schedule pressures will always be imposed on layout activities, but establishment of structured, efficient design processes and procedures can reduce their effect on the quality of the end product.

In today's global market, achieving an efficient circuit board layout cycle requires that accurate, well-defined, and well-controlled information be available to the designer when needed. A complete understanding of the content and structure of the information and how it affects the layout activity will aid in the creation and establishment of balanced processes and procedures. Figure 4.1 shows the normal flow of critical data to the layout process and where and when throughout the layout cycle these data should be defined. In addition to design-related data, the following programmatic information should be provided and evaluated prior to and throughout the design cycle:

- Program requirements
  - Budget/schedule
  - Product performance specification
  - Documentation to be generated and controlled
  - Revision procedures
- Mechanical requirements
  - Definition of all dimensions
  - Mechanical hardware standards
  - PCA installation at the next assembly
  - Environmental and operational constraints
- Electrical requirements
  - Definition of all circuit characteristics
  - Complete parts list
  - Design rules

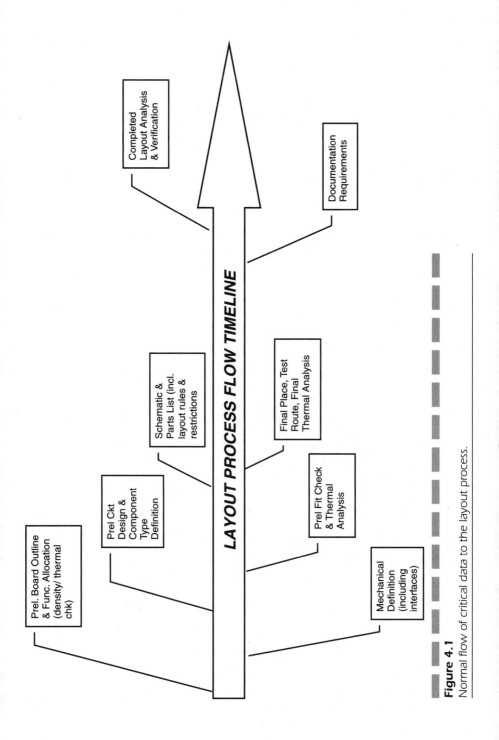

LAYOUT PROCESS FLOW TIMELINE

Prel. Board Outline & Func. Allocation (density/thermal chk)

Prel Ckt Design & Component Type Definition

Schematic & Parts List (incl. layout rules & restrictions)

Completed Layout Analysis & Verification

Documentation Requirements

Final Place, Test Route, Final Thermal Analysis

Prel Fit Check & Thermal Analysis

Mechanical Definition (including interfaces)

Figure 4.1
Normal flow of critical data to the layout process.

103

- ■ Postprocessing requirements
  - ■ Fabrication data
  - ■ Assembly data
  - ■ Testing data
  - ■ Procurement data
- ■ Circuit board layout requirements
  - ■ Layout methods
  - ■ Layout tools
  - ■ Layout support
  - ■ Internal and external interfaces

This chapter will further identify and break down these factors to help identify how to implement effective and efficient circuit board layout processes and procedures.

4.1 Mechanical Definition

One of the key data items that is required in the initial phase of the circuit board layout process is a description of the PCA's mechanical/physical properties and constraints. This information should be clearly and accurately specified, since it is one of the primary constraints on the layout. To effectively verify that the data are correct and complete, a structured mechanical checklist should be used, similar to the one illustrated in Table 4.1.

Mechanical data usually are defined and incorporated into a layout in one of three basic ways:

1. As a manually generated drawing created at either a 2:1 or 4:1 scale

2. As two-dimensional data generated by a computer-aided design (CAD) tool that are used directly as the initial board dimensional baseline

3. As input from a three-dimensional CAD tool that defines all a PCA's mechanical features, including keep-out areas, mounting locations, height restrictions, and all board dimensions

Each of these approaches can provide the required data and documentation necessary during circuit board layout as well as support downstream activities.

TABLE 4.1

Mechanical Description/Constraints Checklist

☐ Board type defined

☐ Board outline dimensions defined

☐ Physical keep-out areas defined

☐ Height restrictions defined

☐ Tooling hole requirements identified

☐ Grid origin dimensioned

☐ Maximum allowable board thickness specified

☐ Maximum number of layers defined

☐ Preassigned layers identified

☐ Core specified (type/material, thickness, layer location)

☐ Laminate materials specified

☐ Component mounting requirements specified (primary/secondary sides)

☐ Connector types and locations defined

☐ Hardware types and locations specified

☐ Solder-mask requirements specified

☐ Solder paste requirements defined for SMT)

☐ Marking requirements defined

☐ Conformal coating requirements specified

☐ Environmental constraints defined

☐ Quality conformance test coupon requirements specified

☐ Panelization configuration defined (if required)

4.1.1 The Board Outline

The board outline definition evolves throughout the circuit board layout cycle as the details of the design are established. It describes all mechanical features and dimensions required for fabricating the bare board and accommodating

- PCA mounting hardware
- Interfacing connectors
- Heat sinks
- Test points
- Large electronic components
- Stiffeners
- Off-grid mounting hole patterns

Figure 4.2 represents a basic board outline. This outline will be used throughout the following sections to help identify the evolution of the PCA's mechanical features.

4.1.2 Keep-Out Areas

Keep-out areas restrict placement of PCA features on a circuit board during layout. They can be defined in many different ways, depending on what the features are and how they are to be controlled. Keep-outs can be unique to single or multiple circuit layers

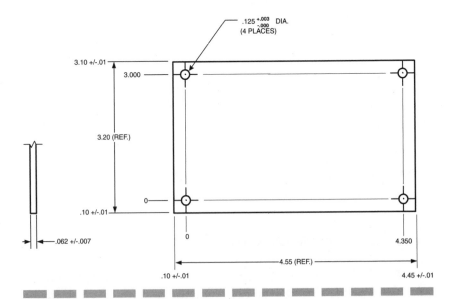

Figure 4.2
Board outline drawing.

(top, bottom, inner, signal only, power/ground, all) and can restrict conductor routing or part placement and/or define height limitations, hole locations, plating, and conformal or solder mask material. In many instances, keep-out areas also may be required to accommodate manufacturing, assembly, or test tools and fixtures. Figure 4.3 identifies some of the various types of restrictions that can be identified and controlled by using keep-out areas.

4.1.3   Height Restrictions

Height restrictions, just as keep-out areas, must be identified and controlled. The basic factors affecting height restrictions are based on

- The space surrounding the PCA when it is installed in its final location
- Access for adjustment and maintenance
- PCA insertion and removal
- Cooling airflow
- Operational environment (susceptibility to shock or vibration)

If a circuit board layout is produced manually, height requirements for all physical parts should be identified either by color-coding methods or by specifying an actual height for each part type. If a printed circuit CAD system is used for circuit board layout, then all height and keep-out restrictions can be defined and embedded within the CAD program for that design. Figure 4.4 identifies an example of how height restrictions are defined.

4.1.4   The Layout Grid

Three main categories of layout grids are used in industry today. They are

1. A *primary placement grid* is normally 0.025 in or larger and is used for locating physical parts, tooling, and datum holes.

2. A *primary routing grid* is normally 0.025 in or less and is used for locating vias, test points, and conductors.

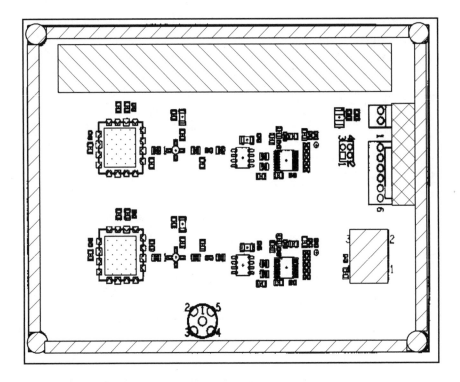

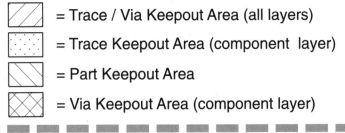

= Trace / Via Keepout Area (all layers)

= Trace Keepout Area (component layer)

= Part Keepout Area

= Via Keepout Area (component layer)

Figure 4.3
Various kinds of keep-out areas on a PCA.

3. A *primary layout area* is gridless (shape-based), is controlled
   within autorouters, and is used for high-density routing of
   conductors.

A grid's origin should be placed at the intersection of a board's
datum lines, usually located at the lower left corner of the layout area.

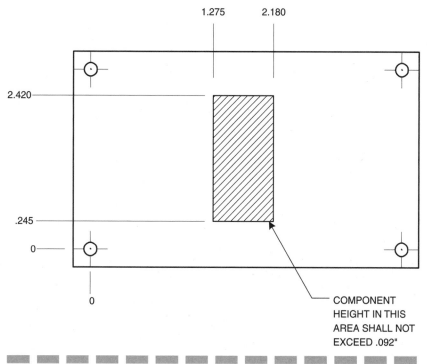

Figure 4.4
Definition of a part height restriction on a PCA.

4.1.5 Tooling Holes

Tooling holes have one essential purpose, which is to act as alignment guides for bare board fabrication, assembly, and test fixturing. In addition, these holes also should be used to locate a board's datum from which all dimensions, component locations, and drill and testing data are referenced. Tooling hole sizes should be selected based on manufacturing guidelines and tooling used, and a minimum clearance should be maintained from component land patterns, probe points, and fiducials.

Three tooling holes, located at the corners of a board at $(0,0)$, $(X,0)$, and $(0,Y)$, are preferred. At a minimum, two tooling holes should be provided, located on a diagonal and spaced as far apart as possible. Figure 4.5 shows typical tooling hole locations.

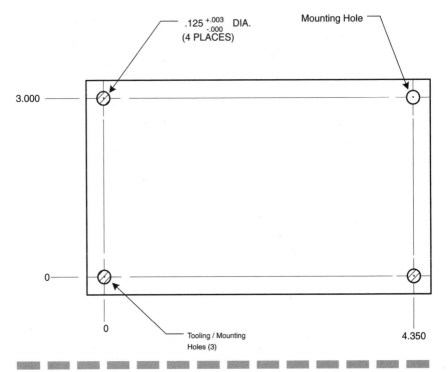

Figure 4.5
Typical tooling hole definition.

4.1.6 Fiducials

The use of fiducial marks facilitates very accurate component placement using automated manufacturing equipment equipped with vision systems. Fiducial marks usually are needed for high-density SMT or mixed plated through-hole (PTH) and SMT designs or for placing devices with high pin counts. Although there are several configurations of fiducial marks, the one most commonly used in the industry is a 0.050-in minimum diameter circle, as shown in Fig. 4.6. There are three different types of fiducial marks:

1. *Component fiducials* are used to increase the placement accuracy of large, multileaded SMT components [leadless and leaded chip carriers (LCCs) with 50 or more pins], fine-pitch (20- and 25-mil) SMT devices, and ball-grid arrays

(BGAs). Two fiducials are required, with one at the centroid of the component footprint and the other placed at the outside corner of the footprint, as shown in Fig, 4.7.

2. *Board fiducials* are required to allow the vision system to identify the board position in the *XY* and theta (rotational) locations. These fiducials should be located in a manner similar to the way tooling holes are located (three preferred, two minimum), as shown in Fig. 4.7.

3. The supplier generally adds *panel-array fiducials* when required to accurately position panelized boards or assemblies.

Regardless of the type, the following general requirements apply to all fiducial marks:

■ The centroid of all fiducial marks should have its *XY* location coordinates located inside the component placement area of the PCA.

■ On very dense PCA designs, fiducial marks may be required to be located underneath through-hole components in order to conserve printed circuit board (PCB) real estate. This

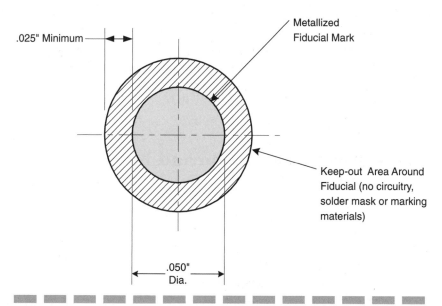

Figure 4.6
Dimensions for commonly used fiducial mark.

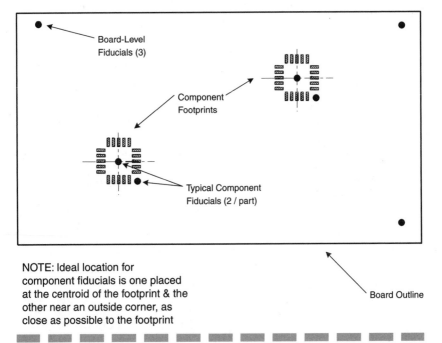

Figure 4.7
Typical placement of board and component fiducials.

requires that all optically located parts be mounted before the through-hole parts are placed on the assembly.

■ The fiducial mark should not be used as a part or extension of a circuit conductor.

4.1.7   Board Construction and Layer Assignments

To establish the optimal layer assignments, the following issues should be evaluated:

■ Overall board thickness limitation

■ Circuit type, performance, and isolation requirements

■ Board material type requirements

■ Thermal, mechanical stress, and layer-balancing considerations

■ Manufacturability

Figure 4.8 identifies a typical board stackup with a cross-reference indicating associated layer assignments.

Individual layer assignments are identified starting from the primary (component or topside) and looking through the design. Each circuit layer is individually identified as layer 1, layer 2,..., layer *n*. The identifier also may include functional information such as layer 2 (ground plane), layer 3 (power plane), and impedance control values.

Once the PCB stackup structure has been defined and the total number of layers identified, the board designer can evaluate and select routing characteristics for each layer. There are five basic options in assigning router characteristics:

1. Primary routing direction vertical

2. Primary routing direction horizontal

3. Primary routing direction diagonal

4. Positive/negative conductor plane layers

5. Pads only, no conductors

Choosing routing directions is based on circuit performance

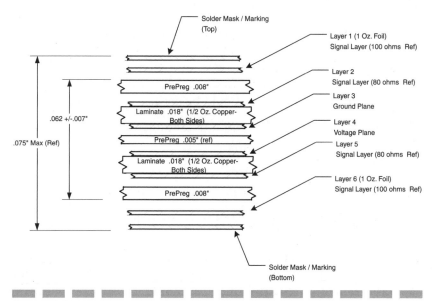

Figure 4.8
A typical board stackup indicating associated layer assignments.

requirements, conductor density, board form factor, types of components used, and board construction. To enhance a board's mechanical strength and prevent warpage and delamination due to mechanical and thermal stresses, it is usually recommended that routing directions be orthogonal to each other on adjacent conductor layers.

4.1.8   Machining Requirements

The most common types of machining processes required on circuit boards are

- Outline definition (routing, edge milling, shearing)
- Drilling
- Slotting
- Counterboring
- Countersinking
- Cutouts
- Surface milling
- Edge chamfering

Figure 4.9 shows how typical machining requirements are specified on a PCB drawing. Some of these machining activities can add significantly to the cost of a circuit board, especially if close tolerances are specified. To help control these costs, the machining requirements should be defined completely, including

- *XYZ* location/identification of all dimensions
- Cost-effective tolerancing
- Use of standard tool sizes

4.1.9   Hardware and Connector Mounting

Commonly, hardware and connector mounting requires additional documentation to specify location and installation requirements. This information should be developed during board design and shown on the layout. Some of the more

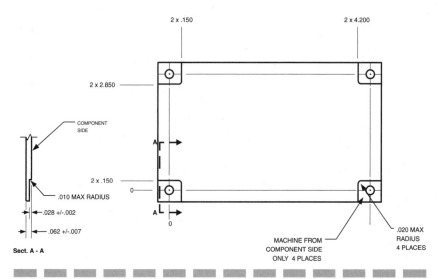

Figure 4.9
Typical board machining requirements.

common types of hardware and connectors used in products today are

- Hardware
  - Screws
  - Wedge locks (for enhanced thermal conduction)
  - Washers
  - Stand-offs
  - Brackets
  - Terminals
  - Rivets
  - Board extractors/ejectors
- Connectors
  - PTH straight, angled
  - Edge
  - Surface mount

The most common design issues to consider when selecting and mounting mechanical hardware and connectors are

- Effect on product performance
- Use of layout real estate

■ Associated material and process costs

■ Identification of process control requirements (location, orientation, insertion, forming, etc.)

■ Rework and repairability

4.1.10 Board Materials

Selection of circuit board materials is a key design decision, and an understanding of the basic characteristics of the types of materials commonly used is very important. The primary factors to consider in evaluating materials are electrical parameters, physical characteristics, cost, and availability. Table 4.2 contains a cross-reference guide to the basic features of various laminate materials.

Many types of copper-clad materials are available, but the ones used most commonly for PCBs FR-4, FR-2/-3, CEM-1, CEM-3, GI, and GT. Table 4.3 identifies these specific laminates along with their construction and common applications. The IPC—Association Connecting Electronics Industries publishes a very comprehensive series of standards that list in detail the properties of all types of laminates, resins, foils, cloths, and processes that are candidates for the manufacture of circuit boards.

4.2 Schematic Generation

Schematic diagrams usually are created using one of the following three methods:

1. Manually drawn, using standard industry circuit symbol templates.

2. By using a computer-aided drafting system.

3. With a computer-aided engineering (CAE) system. CAE tools capture intelligent part and interconnection net data that can be transmitted directly to a PCB computer-aided design (CAD) system.

TABLE 4.2

Basic Features of Commonly Used Laminate Materials

Type	Description	Dielectric Constant (@ 1 Mhz)	Breakdown Voltage (kV)	T_g (°C)	Water Absorption (%)	T_{ce} (in/in/°C) $\times 10^{-6}$	Operating Temperature (°C)	Relative Cost
FR-4	Epoxy-glass	4.8	40	130 (BT epoxy = 180)	0.25	10–12	120 (BT epoxy = 150)	—
FR-2/-3	Phenolic/epoxy-paper	4.4	15–30	110	0.65	12–26	105	<FR-4
CEM-1	Epoxy-paper	4.4	40	—	0.30	11–17	130	<<FR-4
CEM-3	Epoxy-glass	4.6	40	—	0.25	10–15	130	<FR-4
GI	Polyimide-glass	4.8	55	280	0.85	10–12	200	>FR-4
GT	Teflon-glass	2.8	20	—	0.10	10–25	220	>>FR-4

Notes: For dielectric constant, lower is better for high-speed circuitry. For breakdown voltage, higher is better. For T_g (glass transition temperature), higher is better. For water absorption, lower is better. For T_{ce} (thermal coefficient of expansion), lower is better. For operating temperature, higher is better.

TABLE 4.3

Laminate Construction and Common Applications

Type	Construction	Applications
FR-4	Multiple layers of woven glass cloth impregnated with epoxy resin; fire retardant	Widely used in computers, industrial controls, telecommunications, and military/space systems
FR-2	Multiple layers of paper impregnated with flame-retardant phenolic resin	Mainly used in consumer electronics such as games, radios, and calculators
FR-3	Multiple layers of paper impregnated with fire-retardant epoxy resin	Found in consumer products such as computers, TVs, and audio equipment
CEM-1	Glass cloth impregnated with epoxy on outer surfaces of a paper core impregnated with epoxy	Used extensively in industrial electronics, smoke detectors, and for automotive devices
CEM-3	Glass cloth impregnated with epoxy on outer surfaces of a fiberglass core impregnated with epoxy	Used in appliances, automobiles, and commercial communication equipment
GI	Multiple layers of woven glass cloth impregnated with polyimide resin	Used in products exposed to high-temperature environments
GT	Glass fabric base with Teflon resin	Applied where low dielectric constant is needed for high-frequency circuits

CAE systems are now commonly used throughout the electronics industry. These computer-based tools are employed in various stages of the design to model, analyze, and predict the circuit performance of the final physical layout. These tools include

- *Schematic capture systems.* These are used to draw circuit diagrams. They offer capabilities ranging from the simple definition of circuit symbology and interconnection to those capable of performing very substantial error checking.

- *Synthesizers.* These are specialized tools that allow an engineer to specify the logic operations that a design is expected to perform. The synthesizer then extracts the equivalent logic circuit functions from a library and connects them together as specified by the engineer to form a complete schematic.

- *Simulators.* These create computer-based models of a circuit and run them using input test patterns to verify that the circuit will perform its intended function when implemented in hardware.

- *Emulators.* These are collections of programmable logic elements that can be configured to represent almost any kind of logic circuit.

- *Circuit analyzers.* These are used to examine circuits to ensure that they will perform properly over the expected range of timing variations in circuits and component value tolerances. Common examples of circuit analyzers are SPICE and PSPICE. Both build mathematical models of each circuit and perform thousands of complex calculations to predict how the circuit will respond to input signals.

- *Impedance analyzers.* These are used to examine the dielectric and conductor configurations of a design to ensure that the resulting circuit impedance will be within required limits. By making adjustments to various parameters during layout, the final desired impedance can be achieved.

This section will review and identify the processes, methods, and standards used during the schematic generation cycle.

4.2.1 Component Symbology

The American National Standards Institute (ANSI) and the Institute of Electrical and Electronics Engineers (IEEE) are the sources of the two major standards used for identifying and controlling component symbology. These standards describe the size, shape, and drawing options for the graphic symbols used in schematics. The symbols are shorthand methods for graphically representing the functions and interconnections of a circuit. These symbols are correlated with parts lists, technical descriptions, and interconnection instructions by identifying each with a reference designator. Figure 4.10 shows some of the basic schematic symbols.

Information usually associated with the more complex circuit symbols includes

- Pin/terminal numbers
- Pin/terminal function names

- Device type
- Part number
- Reference designation type
- Identification of swappable (interchangeable) functions and pins

4.2.2 Drafting Standards

Schematics should be generated using in-house drafting standards or, if specified, customer standards. If none are identified, then ANSI drafting standards should be used. Some of the common areas that require standardization include

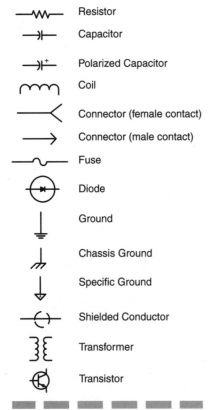

—WW—	Resistor
—)(—	Capacitor
—)(+	Polarized Capacitor
⌒⌒⌒	Coil
—<	Connector (female contact)
—>	Connector (male contact)
—∿—	Fuse
⊕	Diode
⏚	Ground
	Chassis Ground
	Specific Ground
—()—	Shielded Conductor
ЗЄ	Transformer
⊕	Transistor

Figure 4.10
Commonly used schematic symbols.

- Drawing formats
- Title block conventions
- Notes
- Text size, height, and width
- Line weight
- Circuit flow and orientation
- Reference designator assignment
- Interconnection labeling
- Sheet-to-sheet links

4.2.3   Circuit Flow

Schematic symbols and interconnections should be grouped together by function and arranged so that the circuitry flows from left to right and top to bottom (signals in to signals out). Complex circuitry usually will require multiple-sheet schematics. Where possible, component groupings should be placed on individual sheets to minimize the number of interconnections linked to other parts of the schematic.

Terminations linking multipage signals and external PCA connections to other parts of a system should be located at the outer edges of field of the drawing for clarity. Connectors, decoupling capacitors, and spare circuit elements should be located on the last page(s) of a schematic drawing. Where applicable, the schematic diagram also should identify

- Critical/sensitive signals
- Shielding requirements
- Power and ground distribution
- Test point locations

4.2.4   Net Naming Conventions

Adopting a standard set of net naming conventions will aid in streamlining the design process. By using a systematic way of identifying various types of signal and power nets, the opportunity for error during layout, checking, and documentation can be reduced substantially. Duplicate and missing net names are more easily recognized and corrected, and errors made when translating schematic data to the layout environment can be identified readily. When establishing net naming standards, the following should be considered:

- Accommodating multiple variations of power and ground
- Identifying critical and sensitive signals
- Number and types of characters in a net name
- CAE system naming default restrictions

When a CAE system is used to produce schematics, net naming conventions should be structured carefully. Some CAE systems will take control away from the user by forcing default names to be used for certain types of nets. Also, net names may be case-sensitive, and some characters may not be recognized, causing the name to be truncated or changed by substituting a different character. How a specific CAE system deals with net names should be completely understood prior to establishing and applying naming conventions. Figure 4.11 provides some examples of the more common net naming conventions.

4.2.5 Reference Designators

Reference designators usually consist of letters that identify the type of part with which they are associated and consecutive numbers that indicate sequential assignments. For example, C12 defines the twelfth assigned capacitor. Industry standards ANSI Y32.2 and Y32.16 define the letters commonly used for each type of electronic component. Reference designators also may be used to identify where a part is located in a system subdivision by defining the following top-down levels (as shown in Fig. 4.12):

- System
- Set
- Unit number

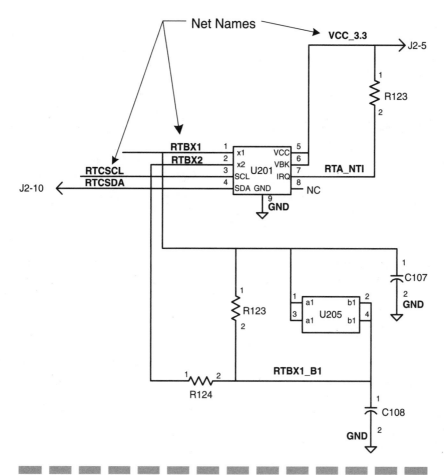

Figure 4.11
Some examples of net names.

- Assembly
- Subassembly
- Basic part

As a circuit is modified, the addition or deletion of parts and their associated reference designations should be controlled in the notes section of the schematic diagram. A note should be provided identifying the highest reference designator number used for each type of part, as well as those numbers which are missing from each sequence. Typical reference designator notes are shown in Fig. 4.13. Once a reference designator has been assigned and then deleted, it should not be used again.

4.3   Component Part Data

Obtaining and validating accurate and complete technical data for each component part type used in a layout is absolutely essential. This descriptive data form the foundation for all PCA design activities and should be checked scrupulously prior to initiating the physical layout activity. Even if a part type already resides in a part library, its description should be verified to ensure that the data exactly characterize the part that actually will be used in the manufactured PCA. This is especially critical when the circuit board layout is generated using a combination of CAE and CAD tools, since an incorrect part parameter will automatically propagate throughout all design and documenta-

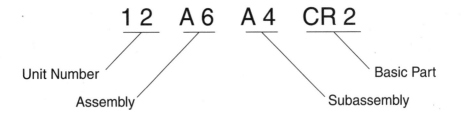

EXAMPLE

1 2   A 6   A 4   CR 2

Unit Number

Assembly

Subassembly

Basic Part

Note: Reference designator number assignment begins below the SET level

Figure 4.12
Full reference designator numbering structure shows where a part is used in a system.

NOTE 6-1

Highest Reference Designations Used
C22, R47, CR17, U33, J6, VR9

NOTE 6-2

Reference Designations Not Used
C3, C9, R17, CR6, U12, J2, VR3

████ ████ ████ ████ ████ ████ ████ ████ ████ ████ ████ ████ ████ ████

Figure 4.13
Typical reference designator utilization notes.

tion operations performed by such systems and corrupt the manufacturing, assembly, and test data they produce.

4.3.1 Part Data Integrity

In order to ensure that a component type is defined correctly in the part library, the following information should be checked:

▧ Manufacturer's data specification sheet that describes the part

▧ Part identification

▧ Physical description

▧ Circuit symbology

▧ Consistency between the data in the circuit symbol library and the physical library

4.3.2 Part Libraries

Part libraries contain information describing each component type used in a design. This information may range from a simple description of a part's body outline and the location of its terminations to complete definition of all its physical and circuit characteristics. Two separate but interrelated libraries are ordinarily maintained for use in designing PCAs:

1. A *physical library* contains all the information required to define the part type, reference designator placeholder, interconnection footprint, part body outline (including height), termination identification and location, placement location origin, and drill, paste, and masking requirements for each part type. Mechanical characteristics such as heat dissipation and weight also may be included.

2. A *circuit symbol library* provides the information required to describe the circuit characteristics of each part type and includes the symbol that defines it functionally and schematically. This may encompass logic properties for digital devices, internal structure, and device termination locations and types (inputs, outputs, power, ground, signals, etc.). Circuit performance parameters, interconnection rules, function and pin swapping, and modeling/simulation data also may be included in a part's library description.

These two libraries may be structured and controlled separately. The physical library is usually created and controlled within a CAD system and contains all the information that is used in the design, fabrication, and assembly of the PCA. The circuit symbol library is usually created and controlled within the computer-aided system used to generate schematics and contains all the information that is used in the schematic and linked to the physical design. This library also provides data for circuit simulation and modeling, as well as for generation of circuit test programming. Figure 4.14 identifies the basic interrelationship of data in these libraries. The linkage between them is best controlled by the use of an internally common part number for both the circuit symbol and the physical shape that represents each unique part.

4.3.3 Part Naming Conventions

There are several industry-accepted standards that should be considered when establishing part naming conventions. They are based on part definitions established by the Joint Electronic Device Engineering Council (JEDEC), a unit of the Electronics Industry Association (EIA).

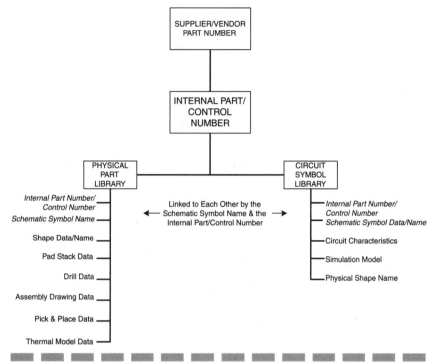

Figure 4.14
Relationship between physical and circuit symbol libraries.

The JEDEC publication JESD1C is an internationally approved document published by the International Electrotechnical Commission (IEC) as IEC Publication 30. Its use facilitates communication among designers, component manufacturers, and board manufacturers by creating a universally recognized acronym system that describes six main component characteristics. These include a part's shape, material, package type, and termination position, form, and count.

The naming approach could use the preceding standard or simply may use a "device type" or "generic type" descriptor. As an example, a standard 14-pin dual in-line integrated circuit (IC) would be identified as an "R-PDIP-T14." This represents a rectangular part (R), made of plastic (P), having dual leads (D), coming from an in-line package (IP), with through-hole leads (T), and a total pin count of 14. The simpler approach would be to just use the identifier "DIP14," which could represent a dual in-line package having 14 pins. The naming conventions used for functional logic part names should incorporate their functional descriptions as well as device or generic type.

4.3.4   Relating Circuit and Physical Definitions

To establish and create definitions relating both circuit and physical properties, the symbol and physical libraries must be linked to each other. The symbol library describes a component's functional characteristics and requirements, such as its primary power/grounds, inputs, outputs, resistance/capacitance values, and other circuit properties. Part terminations defined in the symbol are assigned pin numbers, and the interconnections between them are assigned net names. These same pin numbers are used in the physical part definition and, along with the interconnection data from the schematic, establish a net list that the CAD system uses to define how the parts placed on the layout are to be interconnected. These interconnection nets can be displayed graphically as a "rats nest" that enables a designer to modify the part placement on a layout to optimize routing and enhance circuit performance. Figure 4.15 provides an example of part circuit symbols, physical footprints, and their net list association that defines the rats nest.

Both the physical and circuit symbol libraries can include and provide additional information that supports circuit board layout and downstream manufacturing and assembly/test activities. This additional information may include

- Part numbers (internal/supplier)
- Reference designations
- Quantities
- Location coordinates (drill, testing, placement, etc.)
- Part type performance parameters

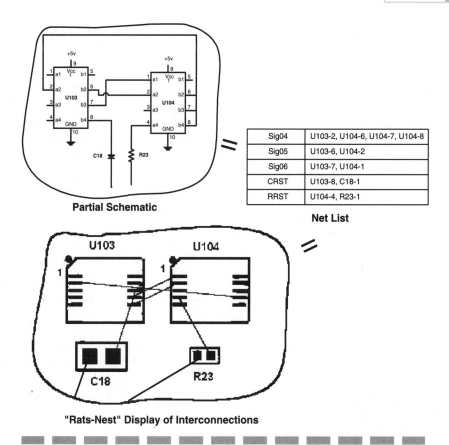

The following tables appear within the figure:

Net List

Sig04	U103-2, U104-6, U104-7, U104-8
Sig05	U103-6, U104-2
Sig06	U103-7, U104-1
CRST	U103-8, C18-1
RRST	U104-4, R23-1

Partial Schematic

"Rats-Nest" Display of Interconnections

Figure 4.15
The rats nets display is established by the relationship between circuit symbols, physical part footprints, and the net list.

4.3.5 Part Footprints

There are two basic types of part footprints, which correspond to the two types of part attachment methods:

1. A through-hole part footprint is based on a part's length, width, and lead location and lead cross-sectional dimensions. Normally, a footprint's pad and plated hole locations are dimensioned to accommodate worst-case tolerance conditions and then adjusted to the nearest intersection of the grid used in the layout process. This

aids in minimizing the cost of special drilling, assembly, test operations, tools, and fixtures associated with producing off-grid circuit features.

2. Surface-mounted part footprints are also defined by a part's dimensional features, but their unique soldering and assembly processes require pad configurations and locations to be designed to ensure that consistent and reliable interconnections are formed. IPC-SM-782, the widely accepted standard for designing surface-mount land patterns, was created in the early nineties to provide the industry with guidelines to help designers define surface-mounted footprints. This document contains basic footprint design formulas, which today have evolved into a set of registered land patterns (RLPs). Using these standard RLPs in a layout provides assurance that the resulting PCA can be assembled cost-effectively and that it will function reliably throughout its intended lifetime.

Figure 4.16 provides examples of both types of part footprints as well as identifying the general guidelines and considerations used during their creation.

When establishing a footprint for a part, constraints other than those imposed by the part itself merit consideration. These additional footprint issues include

- Pad designation for part orientation and termination identification

- Location of footprint origin for placement on the layout

- Pad *XY* location (centerpoint of pin)

- Reference designation accommodation for part/pin identification

- Interconnection escapes

- Solder joint inspectability

- Test access

- Modification and rework capabilities

- Keep-out areas for conductive packages, heat sinks, or thermal pads

LEADED PARTS

Standard Lead   Welded Lead

A = 2 x Lead Dia.(not less than .030", preferably .060')

Basic Formula for Distance Between Holes for Thru Hole Part Leads =
2(A+R) + D + L (Max. Body Length), Rounded up to Next Highest Grid Increment

Lead Dia.	Min. Radius
Up to .030" .030" - .047" Greater than .047	.030" 1.5 x Lead Dia. 2 x Lead Dia.

SURFACE MOUNTED PARTS

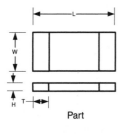

Part

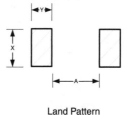

Land Pattern

Basic Formula for Land Pattern:

Land Width (X) = Wmax - Constant
Land Length (Y) = Hmax + Tmax + Constant
Space (A) = Lmax - 2Tmax - Constant

Typical Constant Values	
Rect Chip Parts	.010"
SOICs	.010"
LCCCs	.070" (*)
PLCCs	.030"

(*) With good CTE match

Figure 4.16
Typical guidelines for establishing part footprints.

- Board fabrication restrictions (hole diameters/spacings, annular rings, minimum feature definition, spacings, tolerances)

- Footprint structure (variable layer requirements: inner/outer, signal/power)

132

Chapter Four

■ Solder control (via location, solder dams, paste application)

■ Drilled hole size standardization

4.3.6   Part Mounting and Interconnection

Determining the most effective mounting and termination method for each part type used in a layout requires an awareness of manufacturing and assembly process capabilities that are standard in the printed circuit industry. Design guidelines are available that help identify optimal mounting and termination methods for various types of parts, based on the following parameters:

■ Lead locations and configurations (cross section, length, tolerances)

■ Package configuration (shape, dimensions, tolerances, material)

■ Associated hardware

■ Operating environment (temperature, humidity, shock, vibration, altitude, chemical exposure)

■ Weight of part

■ Heat-removal requirements

■ Access requirements (test, adjustment, visibility)

■ Assembly methods

■ Removal and replacement methods

4.4   Part/Function Placement

Placement of parts and circuit functions is one of the most critical circuit board layout activities. It has a major impact on ease of design implementation, producibility, and ultimately, circuit performance. The factors that should be considered carefully during placement include

■ Functional circuit groupings

■ Circuit signal flow

- Circuit design rules and requirements
- Circuit density
- Connectability
- Environmental effects
- Manufacturability
- Testability
- Accessibility

Throughout this section, how these factors can affect placement will be discussed, along with methods for making the design tradeoffs and decisions that should result in an optimal design.

4.4.1 Part Grouping and Circuit Flow

Establishing effective part groupings and maintaining circuit flow begins by identifying portions of circuitry on a schematic that perform specific functions. Where possible, all components within each circuit function should be kept together, not scattered over the board. The groups of components should then be located on the circuit board such that the functions that interact are adjacent to each other. These guidelines should be balanced against the requirements for furnishing effective thermal paths to cool hot components, mounting of parts to minimize possible damage due to mechanical shock or vibration, providing access to adjustable or display parts, and interfacing with the outside world through connectors.

Additional tradeoffs, involving part orientation for manufacturability, signal flow, and routability concerns, should be made throughout this phase to achieve a design that provides an optimal balance of producibility and performance. Figure 4.17 provides an example of how components in a simple logic diagram can be grouped effectively on a layout.

4.4.2 Restricted Areas

Various types of keep-out areas are established on a layout that may constrain part placement. Some of the more common types of restricted areas affect

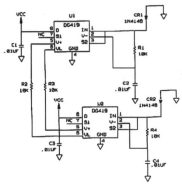

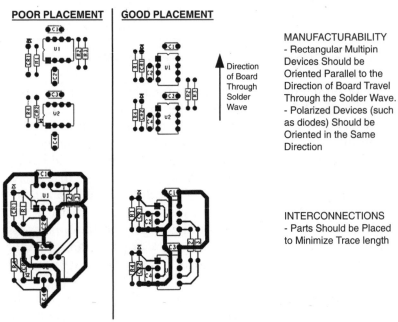

Figure 4.17
Optimizing part placement for manufacturability and circuit performance.

- ■ Parts
  - ■ Height limitations
  - ■ EMI/EMC isolation
  - ■ Thermal interfaces
  - ■ Mounting restrictions (heat sinks, hold-down hardware)
  - ■ Assembly or testing fixtures

- Circuitry
  - Conductors
  - Pads and plated through-holes
  - Vias
  - Power and ground planes
- Marking
  - Visibility
  - Accessibility

CAD tools that are used to produce circuit board layouts can be easily used to identify restricted areas and embed that information within the design database as a layout rule. Some CAD systems can relate this information to characteristics associated with specific parts in the physical library and prevent those part types from being placed in the keep-out areas. Most CAD systems also can automatically control interconnection routing by using preprogrammed keep-in/keep-out rules. If a manual design approach is used, the restricted areas can be defined on the layout and identified in a legend. Some of the more common techniques used include various forms of cross-hatching or color coding.

4.4.3 Circuit Performance and Interactions

It is common on large, multifunction PCAs to have groups of circuitry that interact negatively with each other. To ensure that optimal circuit performance is achieved, the designer should be aware of these interactions and take steps during component placement to avoid them. Some of the more sophisticated CAD tools can help control this type of problem. These systems allow identification of different rule classes that can be structured to prevent them from interacting with each other. Rule classes should mainly define

- *Spacing.* Conductor-to-conductor, conductor-to-land, land-to-land, etc.
- *Isolation.* Circuit type relationships such as digital-to-analog, digital-to-rf, etc.
- *Routing.* Differential pairs, rf signals, controlled impedance, EMC/EMI effects, layer allocations, etc.

Even if the layout tools used do not provide an automated means of preventing interactions that are detrimental to circuit performance, the designer should be able to identify and prevent any potentially damaging interplay between circuitry by manually optimizing the placement. Figure 4.18 identifies some examples for controlling circuit performance and interactions.

4.4.4 Power and Ground Distribution

Awareness of the need to provide effective power and ground distribution for today's high-speed, high-frequency circuits is

External Circuit Trace Isolated From
Other Circuitry by a Metal Shield →

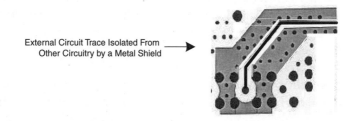

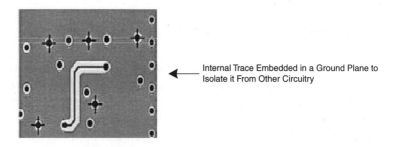

← Internal Trace Embedded in a Ground Plane to
Isolate it From Other Circuitry

High-Speed Data Bus Circuits Routed to
Provide Equal Length Interconnections
& MinimizeTrace Length →

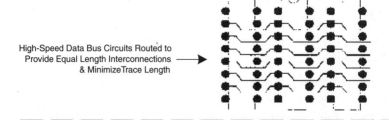

Figure 4.18
Examples of interconnection placement to control circuit performance.

essential. The way power and ground interconnections are laid out physically and the location and placement of bypass capacitors associated with components connected to a power source can have a significant impact on electrical signal integrity and EMI/EMC performance.

To optimize electrical signal integrity, it is desirable to provide dedicated (no signal connections) power and ground planes on separate layers. This is an effective means of providing a relatively low-resistance, low-impedance method for distributing power and ground across the entire area of a PCA. Multiple planes or split planes may be used if there is a need to isolate various voltages (+5 V, +15 V, −15 V, etc.) and grounds [digital, analog, input/output (I/O), etc.]. Figure 4.19 shows an example of how a split power plane can be configured.

Circuit designs with lower operating frequencies and slower rise and fall times generally can be controlled by providing a bus structure for power and ground. These interconnections frequently can share the same layers with signal traces. Figure 4.20 is an example of a routed ground bus.

4.4.5 Trace Density

Trace density not only affects circuit performance but also can have a significant impact on the cost of manufacturing a PCA. Component terminal counts and their placement on a device are the primary factors that influence trace density. Components frequently used in present-day designs have high I/O counts and densities. These require various types of sophisticated footprint designs incorporating many vias in order to provide adequate escape routes for all the device terminations. As the I/O counts grow, trace widths and spacings decrease, escalating trace density and increasing trace lengths. Table 4.4 identifies the impact of increased I/O counts on printed circuit interconnections.

4.4.6 Test Point Requirements

Test point requirements should be considered when placing parts. There are tradeoffs that can be made to determine optimal

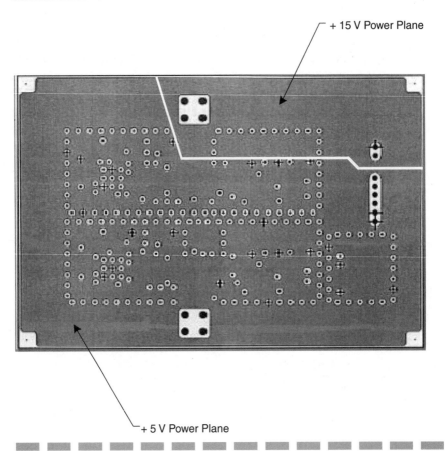

+ 15 V Power Plane

+ 5 V Power Plane

Figure 4.19
Example of a split internal power plane.

test point locations. The factors that are used primarily during this evaluation include

- Type and size of test probes to be used
- Pad sizes and locations that will accommodate the test probes
- Spacing of components to allow test access
- Space requirements and restrictions for test fixtures

Figure 4.21 provides examples of common test point configurations, including probe/pad types, and component and typical pad spacing allowances.

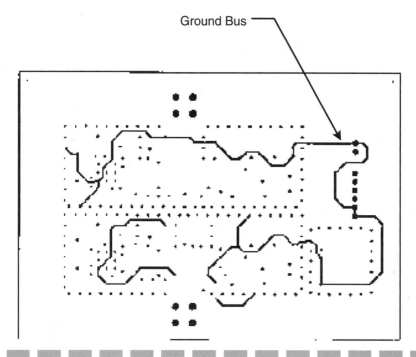

Ground Bus

Figure 4.20
Example of a ground bus routed on a single layer.

4.4.7 Component and Board-Level I/Os

The quantity and location of component and board level I/Os have a major influence on part placement. Component manufacturing capabilities have changed within the last few years. Components can now contain thousands of circuit elements such as digital gates and have I/O counts exceeding 500 pins per device. In order to accomplish this, finer-pitch spacings are being used, resulting in considerably greater interconnection densities on circuit boards. These factors can have a significant impact on part placement, trace routing, assembly, and testing.

As circuit density and complexity increase, the need becomes greater for higher quantities of board-level inputs and outputs. This requirement intensifies placement concerns because I/O connector locations are usually fixed on a board, and many times specific signal and power terminations are prelocated.

TABLE 4.4

Impact of Increasing I/O Count on Interconnection Features

	Increasing I/O Count		
Feature (All Values in Mils)	Std Mixed Technology (Through-Hole and SMT)	Large PGA and BGA* (Microprocessors, etc.)	MCM and Micro BGA
Outer layer traces and spaces	4–5	2.5–3	2–2.5
Inner layer traces and spaces			
1 oz copper	3.25–3.75	3–3.25	3
$1/2$ oz copper	2.5–3	2.5–3	2
Hole diameter— through vias	8–10	8	6
Hole diameter—blind or buried microvias	3–4	2.5–3	2
Pad over drill size	8–12	7–9	6–9
Pad over microvia	6–8	5–6	4
Max. aspect ratio	14	15	16
Max. layer count	30–34	34+	40+
Impedance control tolerances	±8–10%	±6–8%	±5%

*PGA=pin grid array, BGA=ball grid array.

Following the general guideline that devices with I/O connections should be positioned nearest to the connector can result in an unroutable placement. In many designs, additional circuit layers have to be added just to route traces to the I/O connector.

4.4.8 Thermal and Mechanical Vibration/Shock Management

Removal of heat generated by components and possible detrimental effects of operational vibration and shock should be con-

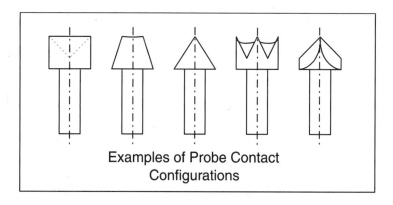

Examples of Probe Contact
Configurations

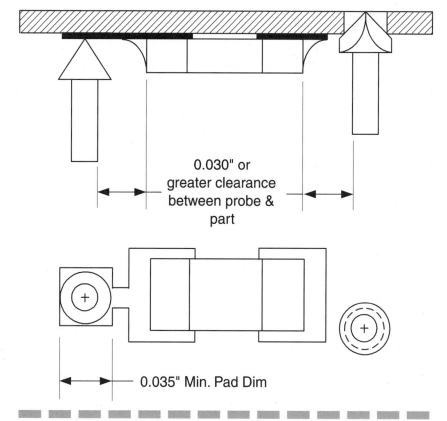

0.030" or
greater clearance
between probe &
part

0.035" Min. Pad Dim

Figure 4.21
Examples of common test point configurations and typical component and pad
spacing requirements.

sidered during part placement. The need for the thermal analysis of dense PCAs is well accepted. To deal with potential failures caused by thermal conditions, the designer should know the location of the largest concentrations of heat so that the part placement can be adjusted to keep part surface and junction temperatures within their specification limits. Computer programs are available that enable designers to model and analyze both static and dynamic thermal conditions on a PCA and aid in identifying effective heat-removal mechanisms. They can perform thermal analyses on large multifunction boards with high heat-dissipation requirements, using part size, configuration, and power dissipation, along with part placement, board geometry, and conductive, convective, and radiant heat-removal mechanisms, to calculate the junction or surface temperature of each device. These thermal modeling programs can help reduce design iterations by predicting board temperatures and gradients and how much component and junction temperatures exceed their respective limits before the first PCA is built. Figure 4.22 is a thermal analysis map of a typical circuit board showing predicted part and circuit board temperatures.

Some of these programs also can be used to optimize part locations and determine their cooling requirements. Increased operational lifetime of PCA circuits generally can be achieved through proper part placement and by judicious selection of materials. Typically, this involves controlling the thermal environment in a populated assembly by the choice of board material and construction and heat sink and heat sink attach materials and use of thermally conductive materials between parts and the board. Using these techniques can result in a substantial reduction in component temperatures and a corresponding increase in reliability.

Computer programs are also available for analysis of physical stress on an assembly and its associated components due to mechanical vibration and shock. They employ sophisticated finite-element modeling (FEM) techniques to graphically display how a circuit board with a specific component layout will react to different kinds of mechanical energy inputs. These data can be used to analyze stresses on connector pins and component leads, allowing predictions to be made on reliability and fatigue life of any PCA component. Figure 4.23 shows a PCA model's reaction to vibration input.

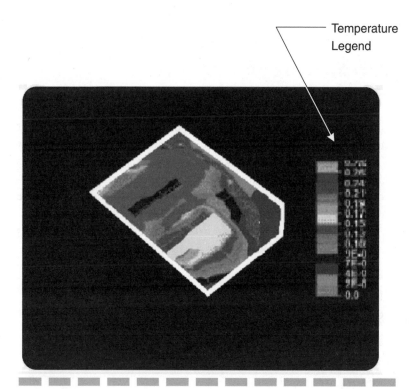

Temperature
Legend

	0.26
	0.24
	0.21
	0.19
	0.17
	0.15
	0.13
	0.10
	8E-0
	7E-0
	4E-0
	2E-0
	0.0

Figure 4.22
Thermal analysis model of a PCA.

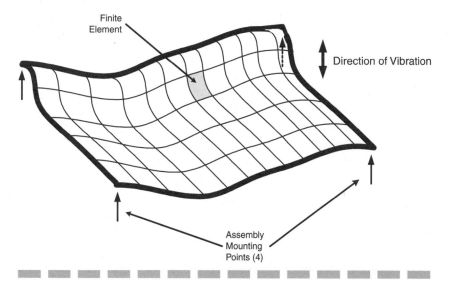

Finite
Element

Direction of Vibration

Assembly
Mounting
Points (4)

Figure 4.23
Finite-element model of populated PCA showing response to vibration input.

4.4.9   Part Spacing and Orientation for Assembly

During placement, emphasis should be put on maintaining part spacing and orientation that will allow efficient and consistent PCA assembly and testing. All downstream processes and tools and fixtures to be used to manufacture a PCA should be identified and their associated requirements assessed early to ensure that part placement will not detrimentally affect the producibility and testability of the end product. The basic processes that can be affected by how parts are located and oriented include

- Part pick and place
- Part attachment
- Soldering
- Inspection
- Testing
- Rework and repair

Selection of the part technology to be used determines the basic PCA manufacturing processes required and therefore what placement guidelines should be applied. Using either surface-mounted or through-hole technologies or a combination of both on a PCA involves use of infrared (IR) reflow, vapor phase, or wave soldering processes each with its own set of process constraints. It is also fairly common to use mixed-technologies due to part availability or circuit requirements. Guidelines for mixed-technology PCAs are complex and may place significant constraints on part placement. Figure 4.24 shows basic configurations used for mixed-technology part placement.

The physical size and quantity of pinouts of parts used also will affect placement because special spacing may be required for

- Nonstandard pick and place tooling
- Prevention of shadowing of nearby joints during soldering
- Additional area for pinout escapes

Figure 4.25 shows some examples of these conditions.

Frequently, the amount of circuitry and components allocated to an assembly may force a designer to use less than optimal part-

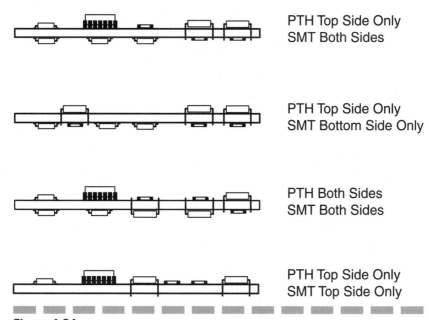

PTH Top Side Only
SMT Both Sides

PTH Top Side Only
SMT Bottom Side Only

PTH Both Sides
SMT Both Sides

PTH Top Side Only
SMT Top Side Only

Figure 4.24
Basic configurations used for mixed-technology part placement.

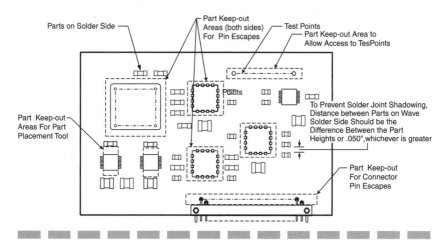

Figure 4.25
Examples of part placement restrictions.

to-part spacings on a layout. At this point, a cost-benefit tradeoff study could be done to determine if producing a single, high-density circuit board is more cost-effective than moving the circuitry to another PCA or creating a new assembly. The factors to be used in making this decision may include

■ Ability of the end system/product to accommodate an additional PCA

■ Cost and schedule impact of creating two less difficult designs versus the additional effort required for a more complex layout

■ Cost of manufacturing two simpler circuit boards versus one board that may require the use of more costly processing for fine-line technology, additional layers, and blind or buried vias

■ Use of automated assembly and test operations versus manual methods for closely spaced parts

■ Performance and reliability effects of using more complex high-density circuitry (circuit integrity, thermal management)

■ Potential repair/rework impacts

4.5 Circuit Routing

A number of PCA design considerations, requirements, and features affect circuit routing. The key areas of concern include

■ Component technology selection
■ Circuit performance
■ Available surface area
■ Interconnection density
■ Routing efficiency
■ Trace/spacing definition
■ Layer structure/balance
■ Environmental/circuit performance accommodation
■ Fabrication/assembly processes
■ Rework/repair considerations.

■ Component I/O quantity and density

■ Testing requirements

■ Controlled signal parameters

■ Routing methods/techniques to be used

The remainder of this section will address some of the more critical factors, along with methods and techniques for identifying, controlling, and sequencing interconnections, layer balancing, and determining trace density and via requirements associated with the routing of circuit board interconnections.

4.5.1   Sequencing Interconnections

Specifying the sequence of interconnections to be made in a circuit net is frequently required to control the effects that are associated with routing critical signal paths. This applies to both digital and analog circuitry but is particularly important to ensure proper operation of complex digital circuits. In order to minimize the effects of circuit coupling and reflections and to maintain the control over signal flow and timing needed for high-speed digital circuitry, it is necessary to arrange the sequence of circuit connections to flow from a net's drivers to its loads and terminations. This arrangement of node-to-node connections in a net is commonly referred to as *signal scheduling* or *sequencing.* Once components are placed on a layout, it becomes possible to determine how to connect the driver to the loads and terminators in order to create the proper signal-transmission lines.

4.5.2   Balancing Layers

For multilayer circuit boards, balancing the amount of conductive material on each layer is needed to prevent the finished board from exceeding its mechanical flatness limits. This requirement could have a detrimental effect on routing efficiency, since the designer may have to put interconnections in less ideal locations to balance the conductor layers. Figure 4.26 illustrates the possible effects of designing a board with unbalanced layers.

A way of dealing with the balancing requirement while maximizing interconnection efficiency is to route traces for best circuit performance and then add dummy traces or conductive material as needed to equalize the layers. An example of this approach is shown in Fig. 4.27.

4.5.3   Interconnection Density

It is important that the interconnection density requirements for a PCA be established early in the layout process. The designer should identify and evaluate the major features that will directly affect a board's circuit conductor capacity and define them so as to ensure a design's routability without placing an undue burden on the producibility of the assembly. How these features are configured will have a major effect on the construction of the PCA (primarily the number and placement of layers) and the rules and procedures that will be used to design it.

A number of mathematical equations have been developed for analyzing a board's connectivity capacity and the connectivity

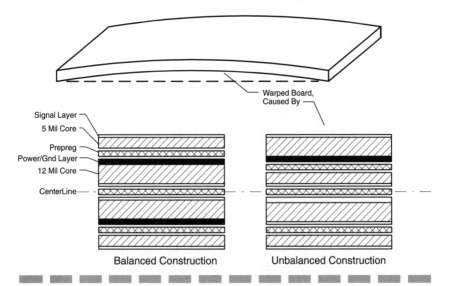

Figure 4.26
The effect of unbalanced construction on board flatness.

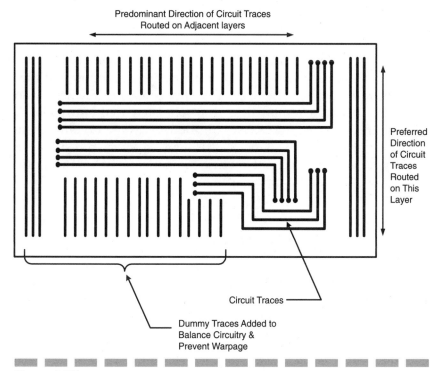

Figure 4.27
Equalizing conductive material to prevent board warpage.

demand placed on it by the circuitry. Unfortunately, because there are usually a large number of unique variables associated with each new circuit design, it is difficult to produce an accurate answer using this approach. An alternate method is to assess a circuit's interconnection-related characteristics and adjust the physical features of the printed board layout to satisfy them. These characteristics may include

■ Component technology

■ Component spacing

■ Component termination density

■ Number and complexity of circuit nets

■ Circuit grouping and isolation requirements

■ Voltage/current requirements

■ Thermal/vibration requirements

■ Testing requirements

Component I/O density is probably the single most important factor. It is basically driven by device pin counts and their pitch or spacing. The major interconnection requirements associated with a circuit design, when identified and properly defined, will aid in the determination of layout grid dimensions, routing channels, conductor widths and spacings, pad sizing, via types, and placement of specific components. Table 4.5 identifies how these issues define routability.

4.6   Circuit Performance

A PCA's overall value is determined by the performance of the circuit it contains. To ensure that a product fulfills its functional requirements, the features of a layout that have the greatest

TABLE 4.5

Impact of Increasing I/O Count on Board Routability

	Increasing I/O Count		
Feature	Std Mixed Technology (Through-Hole and SMT)	Large PGA and BGA (Microprocessors, etc.)	MCM and Micro BGA
Average no. of pins/in^2 of board area	60–80	100–200	250 or more
Average trace length/ interconnection	2.5 in	3 in	3.5 in
Average wiring demand (in/in^2)	60	120	>160
Wiring efficiency (use of available routing area)	Up to 50%	Up to 65% (with blind/buried vias)	Up to 75% (with blind/buried microvias)
Max. traces/0.050-in grid	2	3	4

influence on circuit performance should be identified and controlled. This section discusses and reviews these factors and provides some guidelines, methods, and techniques that will aid in implementation. This section also discusses the impact of layout rules, timing issues, signal integrity, transmission-line effects, impedance controls, and EMC/EMI shielding requirements on circuit performance.

4.6.1   Layout Rules

Producing an effective design starts with the establishment of a basic set of layout rules. These rules should balance and account for the needs of all design and manufacturing process requirements. They will vary depending on the complexity of the design as well as its performance objectives. To ensure that layout rules are established properly and completely is a task that should involve all members of the design and manufacturing team. The major issues to be considered during the layout rules development process include

- Power distribution and coupling
- Critical signal paths
- IR drops in signal and power conductors
- Impedance control
- Pad/land geometries
- Conductor widths /spacings
- Via types
- Plated/unplated hole sizes
- Part placement constraints
- Layer assignments and routing constraints
- Dielectric material and thickness
- Conductor material and thickness
- Heat-removal paths
- Part mounting constraints
- Fabrication/assembly/test requirements

Once layout rules have been defined, they should be documented and followed throughout the design cycle. Layout rules can be incorporated into most circuit board CAD systems, providing the capability to automatically prevent them from being violated or at least flagging potential violations during the layout process. Even with these controls in place, manual intervention by the designer sometimes can occur. It is important that deviations from predetermined layout rules be identified and documented.

When establishing layout rules, consideration should be given to how they will constrain the layout. Even the most powerful CAD system may not be able to produce a cost-effective design because of the limits placed on it by the embedded layout rules. A designer should understand how the CAD system being used functions and accommodate those capabilities when developing the design rules for a PCA.

4.6.2 Timing

Understanding the effects that conductor lengths have on signal timing is essential to meeting overall circuit performance. Signals that have timing constraints, particularly digital clock lines, should be identified early in the layout process and considered during component placement. Timing is mainly affected by the length of conductors between devices. Figure 4.28 shows a typical timing diagram. High-speed circuitry, with very short signal rise and fall times, is particularly sensitive to timing variations. If the time of travel for signal approaches one-half the rise or fall time, circuit performance will be degraded, and the circuit may cease to function properly. Figure 4.29 shows how signal propagation delays vary with different substrate dielectric properties.

4.6.3 Signal Integrity

Maintaining signal integrity has become a major challenge when designing PCAs containing high-speed circuitry. In the past, signal integrity was not a significant concern because clock rates

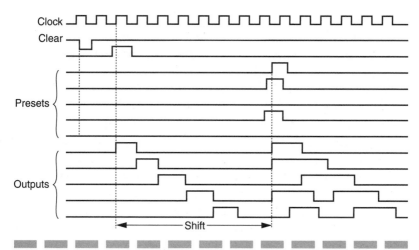

Figure 4.28
Typical timing diagram (shift register).

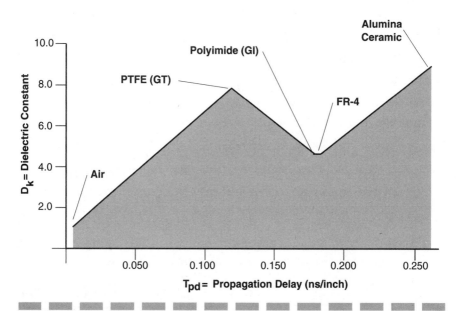

Figure 4.29
Signal propagation delays for materials with different dielectric properties.

and signal edges were relatively slow. Recently, however, circuits with clock rates exceeding 100 MHz and less than 1-ns signal rise times are commonly used in industrial, commercial, and consumer electronic products, and their performance can easily become degraded if circuit board interconnections are not configured properly.

Signal integrity basically involves preserving the intended form and purity of circuit pulses. As circuit speed and edge rates increase, board interconnections play a significant role in determining circuit performance. Because of the effects of frequency and voltage/current, a variety of distorting conditions can exist, determined primarily by the way interconnections are defined, that may have a negative effect on signal integrity. These include

- Interconnection delay
- Ringing
- Signal reflections
- Crosstalk
- Simultaneous switching noise
- Signal skew
- EMI effects

Discontinuities in pad and conductor geometries have a major effect on signal integrity. This can be minimized by keeping conductor paths as short as possible, as well as by equalizing conductor lengths in a net. Crosstalk and noise are other sources of signal degradation. These conditions are prevalent in high-density designs having areas of closely spaced conductors. Crosstalk may be reduced by limiting parallel conductor paths on the same signal layer or on adjacent signal layers, as well as by ensuring that sensitive data lines are adjacent to a signal ground line or plane. Reducing impedance mismatch between component terminations and their conductor paths also can decrease signal-line discontinuities. In some instances, resistive terminations may have to be added to provide characteristic impedance matching, especially if long traces are involved.

4.6.4 Impedance Control and Transmission-Line Effects

The characteristic impedance of conductor paths for high-speed circuitry can be defined by designing them as transmission lines. Transmission lines consist of conductors spaced above, below, or between metal reference planes (usually power or ground planes) in a multilayer construction. The value, accuracy, and consistency of the resulting impedance of the transmission line depend on close dimensional control over conductor width and thickness, the thickness of the dielectric material between the conductor and the metal planes, and the dielectric constant of that material. The basic types of transmission-line constructions are

- *Open microstrip.* The conductor is located on an external surface and separated from a reference plane by a single thickness of dielectric material.
- *Embedded microstrip.* The conductor is bounded on all sides by dielectric material and separated from a single reference plane.
- *Balanced stripline.* Similar in construction to an embedded microstrip, except that the conductor is separated equidistantly from two reference planes.
- *Unbalanced or asymmetric stripline.* Similar in construction to an embedded microstrip, except that the conductor is separated unequal distances from two reference planes.

Figure 4.30 shows examples of these transmission-line structures.
Impedance control requirements should be specified prior to starting physical layout activities. They are usually established as part of the circuit design activity through modeling, simulation, and signal integrity analysis. Which of the preceding types of transmission-line constructions are used on a circuit board and their physical configuration are defined by the board designer in concert with the circuit engineer. Since selection of board construction and choice of types and thicknesses of various dielectric materials, along with tight control of dimensional tolerances, are also critical factors, the board fabrication operation also should participate in these design decisions. Industry

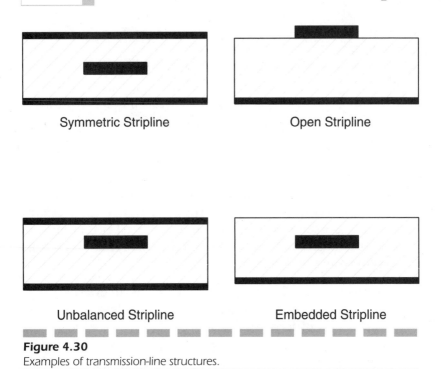

Figure 4.30
Examples of transmission-line structures.

guidelines are available that provide formulas for computing the dimensional, conductive, and dielectric parameters required for establishing conductor paths with controlled impedance values.

4.6.5 Electromagnetic Interference (EMI) and Electromagnetic Compatibility (EMC)

Understanding the effects associated with EMI and EMC and how to control them is imperative in the design of any electronic hardware. EMI involves unintended transmission, by radiation or conduction, of electromagnetic energy that disrupts the performance of the receiving circuitry. It usually refers to signals in the rf range (10 kHz to 100 GHz), but EMI can occur at all frequencies. EMC is the ability of circuitry to exist in an electromagnetic environment without suffering performance degradation.

Almost any type of circuitry can be either a source or receiver of electrical noise. When both these conditions exist in the same system or on the same PCA, the potential for conflicts can present significant and challenging layout problems. To ensure that internal and external EMI sources will not impair circuit performance, both circuit transmitters and those sensitive to electromagnetic transmissions should be analyzed, and steps should be taken to avoid possible performance corruption.

Methods of mitigating the effects of EMI on circuit board layouts include

- Use of external ground planes

- Embedding noisy signals between reference planes

- Isolating/partitioning analog and digital circuitry

- Separating parallel signal traces by a minimum of twice the trace width spacing

- Routing power and ground traces in parallel to minimize current variations

- Locating components as close as possible to ground connections to minimize ground and signal loops

- Controlling trace impedance

- Providing metal shields around noisy components

Allowable emission levels generated by electronic products are regulated by national and international organizations. In the United States, the Federal Communications Commission (FCC) specifies allowable emission levels. Internationally, most countries have established a similar regulatory agency. Technical standards and operational requirements for electromagnetic emissions of products used in the United States can be found in the *Code of the Federal Regulations* (CFR), Title 47, Parts 15, 18, and 68.

4.7 Routing the Interconnections

The effects of the requirements and constraints described in the previous sections of this chapter on the actual routing of a printed circuit board may seem overwhelming. This section describes some of the methods and techniques that will address this concern and help simplify the routing process.

Prior to starting the actual routing process, the major design factors described earlier should have been identified and defined. These factors, which are derived from the overall PCA performance requirements, should include

■ Component placement

■ Voltage/current requirements

■ Circuit performance requirements

■ Board material and construction

■ Fabrication, assembly, and testing constraints

■ Operational requirements

Having these factors previously established allows the designer to remain focused on the issues pertaining to the routing process.

4.7.1 Conductor Widths and Spacing

When defining conductor widths and spacings for a layout, the key factors that should be considered are

■ Trace density and routability (see Sec. 4.5.3)

■ Circuit performance requirements (see Sec. 4.6)

■ Board producibility

Conductor width determines the electrical characteristics of an interconnection, mainly by limiting current-carrying capacity due to resistive loss. Graphs are available that show standard current-carrying capacities for various trace widths and thicknesses. Most allow a margin of safety that takes tolerances associated with the board fabrication process into consideration. Figure 4.31 shows a current-carrying capacity graph.

Although nominal conductor widths for a design are defined by circuit pattern artwork, their finished dimensions may vary depending on the process used to fabricate the circuit board. A 0.010-in wide conductor on an artwork can vary in cross-sectional area as much as 15 percent of nominal on the finished board. It is therefore important that a note be included on every board detail drawing specifying minimum allowable conductor width and thickness and allowable deviations from the artwork. *Con-*

ductor spacing is defined as the distance between adjacent edges (not centerlines) of isolated conductive patterns on a common layer. Spacing is measured at the closest point on adjacent conductors, as shown in Fig. 4.32. Again, spacing may deviate from that shown on the artwork, and minimum allowable spacing should be specified on the board detail drawing.

4.7.2 Layer Utilization

Layer assignment is critical to circuit performance. In addition to the physical issues described in Secs. 4.1.7 and 4.5.2, circuit performance requirements, as discussed in Sec. 4.6, should be considered. Significant factors to be addressed when determining layer use include

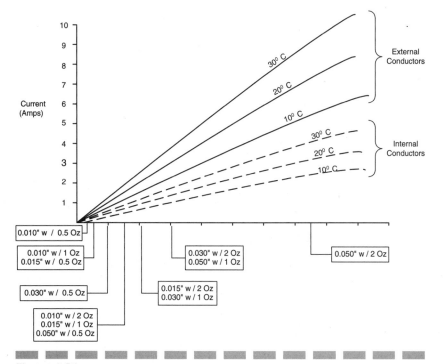

Figure 4.31
Chart shows nominal temperature rises above ambient versus current for various conductor widths and thicknesses.

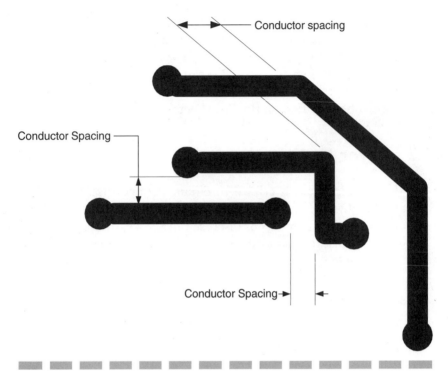

Figure 4.32
Examples of conductor spacing.

- Circuit isolation
- Voltage and ground distribution
- Trace impedance
- Circuit shielding
- Via types required

How these factors affect layer utilization is identified in Figs. 4.33 through 4.36.

4.7.3 Via Types

A *via* is a metallized hole that provides a conductive path between circuit layers on a circuit board. Three basic types of vias are used in circuit board designs. They are shown in Fig. 4.37.

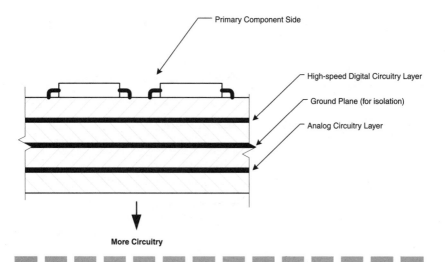

Figure 4.33
Example of use of circuit layers for isolation.

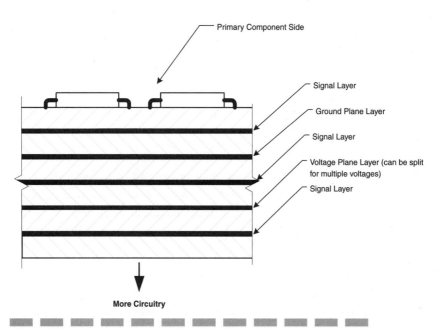

Figure 4.34
Example of use of circuit layers to distribute voltage and ground.

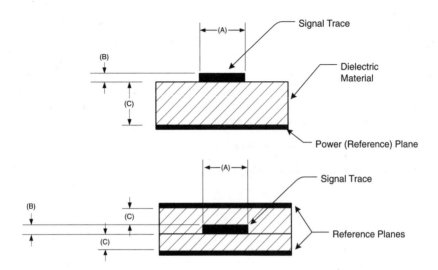

The Characteristic Impedance Value of a Trace is Primarily Based on its Width (A), Thickness (B), the Dielectric Thickness (C) Between the Conductor & the Reference Plane(s), & the Electrical Properties of the Dielectric Material

The Impedance Value Tolerance % of a Trace is Mainly Defined by the Conductor Width (A) Tolerance %, Which should = the Dielectric Thickness (C) Tolerance %

As Circuit Densities & Circuit Speeds Increase Tighter Tolerances are Required for Controlled Impedence Lines:

Present	Near Future
± 8% - ±10%	± 5%

Figure 4.35
Examples layer utilization for controlled impedance circuit structures.

1. A *through via* provides a connection between external and internal layer circuitry. This via type is the most commonly used and least costly to implement.

2. A *blind via* provides a connection between an external layer and one or more internal layers. This via type is used mainly when isolating critical circuitry and is more costly to fabricate.

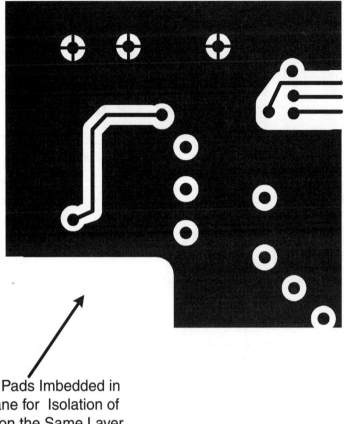

Traces & Pads Imbedded in
Metal Plane for Isolation of
Circuitry on the Same Layer.

Figure 4.36
Isolation of circuitry routed on a common layer.

3. A *buried via* provides a connection between internal layers
only. This type of via is usually used in very dense PCBs and
is the most costly to fabricate. It can be formed using a variety
of processes, including mechanical drilling, laser drilling, or
chemical etching.

4.7.4 Autorouting and Manual Routing

The methods used to route circuit board interconnections have
changed drastically over the past 10 to 15 years. This is primarily

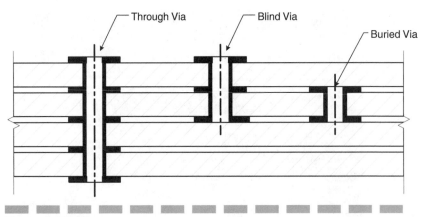

Figure 4.37
Basic types of vias.

due to increased circuit complexity and density, which was brought about by

- The evolution of large, complex integrated components with high pin counts
- The availability of high-performance board laminate materials
- The development of PCB fabrication processes and materials needed to reliably produce high-density circuitry
- The development of highly automated assembly and test equipment
- Improvements in CAD tool capabilities

Working with Bishop tape and Exacto knives, designers once produced printed circuit designs that were considered works of art. This era of manual routing, when a designer exercised total control over the location of every part, pad, and conductor on a circuit board, gradually disappeared with the advent of printed circuit CAD systems and declined further with the appearance and increased use of autorouters.

Autorouters find paths for required interconnections by following a prescribed set of wiring rules and do so much more quickly that a designer can manually. However, an autorouter is limited by the level of complexity of the rules it can comply with and is therefore unable to control the shape and location of every conductor on a layout. Also, wiring rule constraints frequently will prevent an autorouter from successfully completing all the interconnections, and significant manual editing of the layout will be required to finish a design. Autorouters are notoriously poor at considering manufacturing requirements, and in some instances, a large amount of manual cleanup is needed to enhance the producibility of a layout.

While pressure to get a complex design to market in the shortest period of time justifies the use of an autorouter, its shortcomings should be understood and accommodated by the designer. Many of today's PCA designers know how to use autorouters only from the standpoint of following a simple, standard procedure: Place components; set a few simple parameters for number of layers, routing directions, and similar basic requirements; and then "hit the route button." When routing is completed, they may spend a little time on cleanup and then generate plot files for board fabrication. The results may be quick but are not usually optimized for circuit performance or PCA producibility.

There is, however, a sequence of actions that can be used even with the most basic autorouters that will enable a designer to quickly produce a completed route that requires minimal cleanup. The steps are

1. Use the autorouter to evaluate the component placement. Run test routes with minimal constraints to identify interconnection problem areas.

2. Adjust placement to improve the route, and perform another test run. This is to be considered a "quick and dirty" phase that is not concerned with via minimization or any kind of cleanup. The idea is to achieve the highest possible percentage of completions, comply with circuit performance requirements, and enhance board producibility while optimizing component placement. Steps 1 and 2 may be repeated as often as necessary.

3. Once an optimal component placement is achieved, critical signals should be rerouted interactively, if required, and locked so that the autorouter will not be able to move or rip them up during any rerouting.

4. Perform a final autoroute "for the money."

4.7.5 Controlling Trace Placement and Lengths

Identifying critical circuits and the effects that trace placement and lengths have on their performance will aid the designer in determining the effort to be expended in manually routing and locking interconnections prior to autorouting. The designer should appreciate that fixing the location of traces prior to autorouting the rest of the interconnections may significantly limit a routing program's ability to complete the hookup successfully. When assessing the need for preplacing interconnections, the following items should be evaluated:

- Signal requirements (clock, rf, impedance, differential pairs, etc.)
- Component placement (optimize short routes, decoupling, power/ground connections, etc.)
- Wiring density (trace width/spacing, routing channels, etc.)

In addition to these factors, trace placement relationships with component and via lands also should be considered.

4.7.6 Editing

As a design evolves through the design cycle, there are several places where editing of the layout can be beneficial. Editing usually should be done during part placement and circuit routing. Editing that commonly occurs during these phases includes

- Part placement
  - Moving, reorienting components
  - Swapping internal circuit functions and device pins
  - Changing component pad stacks
- Routing
  - Rerouting traces
  - Modifying trace widths and spacings

Edits can be implemented locally or globally. An example of this would be if during the routing phase 8-mil traces were used and on route completion it was apparent that there was room for 10-mil traces. A simple global command could implement this change, resulting in a major layout improvement. Most CAD systems also will allow local deviations from the preset wiring rules so that conductor widths may be expanded selectively where possible. Regardless of the type of editing implemented, current CAD tools have the capability to identify, monitor, and control edits as they occur to ensure that design rules are not being violated.

4.7.7 Power and Ground Interconnections, Planes, and Embedded Circuitry

Implementation of power and ground interconnections, reference plane structures, and embedded circuitry is determined primarily by circuit performance requirements. There are two basic methods for providing power and ground interconnects:

- Individually routed traces
- Solid reference planes on individual layers

Individually routed power and ground lines are commonly used for low-speed analog and digital circuitry. As higher-speed circuitry introduces the need for maintaining signal integrity and improving noise immunity, the use of full power and ground planes on a multilayer board structure becomes desirable. If an individual PCA incorporates both analog and digital functions or a mixture of device families operating at different signal speeds, their circuitry should be partitioned into separate areas so that isolated or split power and ground planes can be allocated for each type of circuitry. To perform their intended function properly, these planes should be continuous. Embedding signal or power traces in a plane or creating large etched-out areas results in discontinuities that may create signal and ground current loops and degrade overall circuit performance.

When solid-plane layers are provided, circuit connections to them should be made by a plated hole surrounded by a thermal isolation pad at the layer. The purpose of this type of pad is to prevent formation of a cold solder joint caused by the heat-sinking effect of the metal plane, as well as to enhance repairability. They help to retain heat applied when remelting a solder joint during a rework or repair activity. These pads may be constructed as shown in Fig. 4.38.

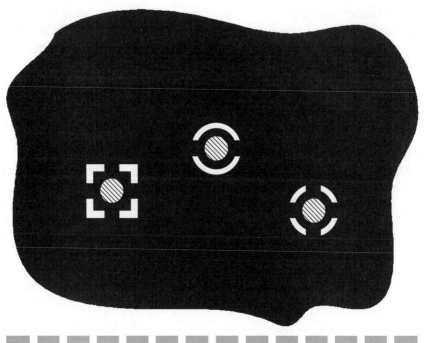

Figure 4.38
Typical thermal isolation pad configurations on a metal plane.

4.7.8   Surface and Interlayer Routing

For a variety of reasons, the physical features and characteristics and locations of conductors on external layers of a circuit board usually will be unlike those of the equivalent ones placed on internal layers. These variables are mainly influenced by a board's construction features, the parameters of the laminates used, and the processes and materials required to produce the board. These include

■ Characteristic impedance

■ Heat rise

■ Copper weight (thickness)

■ Conductor width and spacing

■ Etchback factors

■ Plating

■ Solder mask use

■ Copper balancing between layers

Table 4.6 lists the differences between external and internal conductors that are caused by these influences

Factors that commonly influence both external and internal routing include

■ Allowable stub lengths with and without terminators. Trace stubs may cause signal reflections and also can act as unwanted signal antennas.

■ Balancing conductor signal delays by adjusting trace lengths, especially for the arms of a T connection.

■ Right-angle turns and T's. Signals conducted by a straight trace are relatively clean. Traces routed at right angles or using T's frequently will cause signal degradation, and the corners will be electromagnetic radiators. Where possible, signal turn angles should be 135 degrees or greater.

4.7.9   Interconnecting Large Devices

A number of unique design issues are associated with laying out boards that incorporate large, multifunctional devices. The evolution of increasingly complex electronic devices and their widespread use have increased dramatically over the past few years and will continue to do so. The features inherent in these types of components that challenge a designer are

■ Very high I/O pin counts (500 and climbing)

■ Closer pin spacings (less than 50 mils and shrinking)

■ Pin array packages (PGA, BGA)

■ High heat concentration

■ Test access

■ Quality assurance after assembly (cleaning and inspection)

These factors contribute to a significant increase in both design complexity and the difficulty of manufacturing and assembling the resulting PCAs. As device I/O counts increase, the area required to place pin escapes drastically affects a board's

TABLE 4.6

Differences in Physical Features and Characteristics of Internal and External Conductors

Conductor Feature	External	Internal
Characteristic impedance control	±8–10% (surface trace impedance defined by controlled dielectric material on one side only)	±5–8% (buried traces surrounded by material with controlled dielectric properties)
Heat rise	Lower—Heat can be dissipated into surrounding air.	Higher—Approximately twice cross-section area required for same current and allowable temp. rise
Copper weight	Usually 1 oz copper	Usually $^1/_2$ oz copper
Width	5–10 mils (1 oz)	2.5–3 mils ($^1/_2$ oz)
Spacing (up to 100 V)	5 mils (coated)	4 mils
Etchback (trace width reduction)	2–3 mils (1 oz copper)	1–1.5 mils ($^1/_2$ oz copper)
Plating	Required for solderability and corrosion prevention	None usually required
Solder mask	Usually needed for manufacturability	Not required
Copper balance between layers	Unbalanced external layers have greater influence on board warpage	Layers closer to center of board cross section have less influence on board warpage

overall routability. This problem worsens as the pin-to-pin pitch decreases. Smaller via lands and plated holes result in minimal annular rings, thinner traces, and smaller spaces, which may have a detrimental effect on signal integrity. Also, more routing layers are needed, tending to create marginal hole/board aspect ratios.

All these potential effects should be analyzed, and if appropriate, consideration should be given to creating an additional PCA rather than attempting to develop a single design that may be very costly to produce and which may exhibit marginal performance.

4.7.10 Producibility

A design that cannot be manufactured consistently within its cost and schedule constraints is a poor design. During the layout process, producibility considerations should address three major process areas: board fabrication, component assembly, and performance testing. The general factors that affect each of these areas include

1. Board fabrication
   - *Material.* Cost and availability
   - *Conductors.* Width and spacing
   - *Copper balancing.* Weight and area
   - *Annular ring.* Aspect ratios
   - *Plating.* Types and locations
   - *Drilling.* Sizes and quantities
   - *Routing.* Types, radius, and locations
   - *Tooling.* Soldermasks and stencils

2. Assembly
   - *Mounting.* Hardware, types, and sizes
   - *Components.* Types, sizes, and orientations
   - *Equipment.* Part placement and mechanical processes (swaging/staking)
   - *Placement fiducials.* Types, sizes, and locations
   - *Assembly.* tools and fixtures
   - *Rework/Repair.* Equipment and processes
   - *Identification.* Marking types and locations

3. Testing
   - *Equipment.* Bare board, in-circuit, functional
   - *Fixtures.* Single-sided, double-sided, clam shell
   - *Test probes.* Types, sizes, and locations
   - *Clearances.* Component, card edge, and test point locations

4.8   Checking and Analyzing Routed Results

After interconnection routing has been completed, the layout should be verified (checked) to confirm that the design rules and requirements that were defined initially have been implemented correctly and completely. Areas to check include

- 100 percent compliance of circuit interconnections with the schematic
- Conductor width and spacing
- Sensitive or critical signal routing
- Signal integrity
- Physical component land and pad spacing
- Test point access

All these factors should be analyzed and verified regardless of whether a manual or automated routing technique was used. Using CAD tools allows the routing verification process to be significantly shortened through the use of system-generated reports that can provide a list of various interconnection parameters and possible rule violations. The remainder of this section will review processes, methods, and techniques that may be used to analyze and verify circuit performance, design rules, signal routing decisions, and test coverage issues.

4.8.1 Circuit Performance Analysis

After all signal connections have been routed, the physical shape, length, and location of each trace will have been defined, as well as its individual net relationship. These physical data can serve as input to computer programs that will analyze the effects of the physical configuration and placement of the conductors on critical circuit performance parameters such as signal timing and transmission-line characteristics and identify potential anomalies. Normally, problems that are detected during this verification process can be eliminated by manually editing the layout.

Following this, the completed route should be checked against the schematic to ensure that there are no discrepancies. A check should then be performed on the master artwork data to verify compliance with the line width and spacing requirements. The artwork also should be checked to make sure that there will not be any solder mask or markings overlap onto component pads or lands and that traces and other features that should be protected from solder will be covered properly by solder mask material.

4.8.2 Revalidating Design Rules

It is important that the designer assess how application of the design rules has affected the layout after circuit interconnection routing is completed. This review should verify that the following basic design constraints have been accommodated properly:

- Conductor width and spacing requirements
- Component pad and land sizes and spacing requirements
- Critical signal routing requirements
- Component placement, marking, and orientation requirements
- Layer assignments
- Net sequencing
- Signal isolation and routing requirements
- Power and ground distribution requirements
- Conductor length restrictions
- Stubbing limitations

Each of these factors should be verified to ensure PCA performance integrity. The more capable autorouting programs can effectively control certain types of signals that may require matched lengths or need to be isolated to reduce crosstalk. However, it may be more effective to preplace critical or sensitive traces prior to initiating the router, since some routing routines, when faced with conflicting rules, may create less than optimal conductor paths.

In addition to validating the implementation of design rules, the verification process also should assess their detrimental consequences. Such conditions will not be identified as violations by a CAD system's design rule checker because they satisfy the basic layout constraints and are therefore of no concern to the router.

An example of this might be the rule for conductor to pad spacing that has been set for 0.010 in due to the concern for routability because of the density of circuitry. In several areas this spacing could be increased to 0.050 in, which may reduce unwanted signal interaction and improve the board's producibility. The spacing, however, will remain at 0.010 in because of the preset design rule.

This occurs quite frequently during autorouting and is mainly due to the software algorithms these programs use to position the traces. It is important that the designer examine the conductor pattern produced by the router and, if needed, manually revise it to improve circuit performance and producibility of the design. Many of today's CAD systems include editing routines, such as grouping, push and shove, or sliding, that enable a designer to greatly improve a previously routed layout with minimal effort.

Design Quality

Throughout the design process for printed circuit assemblies (PCAs), and especially at key or critical points in the process, specific procedures and activities should be undertaken to verify the correctness and quality of the design. These activities include performing design reviews, checking the accuracy of the design and its associated data and documentation, and evaluating the producibility, testability, and projected reliability of the product.

Frequently, budget and schedule constraints put pressure on the designer to minimize design quality verification efforts. Unfortunately, the complexity of today's PCAs almost guarantees that a less than scrupulous review, check, and analysis will result in a deficient or defective design and in many cases require that the PCA be redesigned.

It is important that PCA quality verification procedures are standardized and the procedures documented. Doing so establishes a disciplined methodology that ensures that the proper things were assessed in the correct sequence using the right procedure.

Documenting standardized verification procedures and following them provide many benefits. One of the main advantages is that the standard process can be evaluated and improved through lessons learned from its prior use. It also enables designers and engineers to assess and verify each other's work using a common set of standards and requirements. Evolving from standard procedures can be a common set of tools that improve the effectiveness and efficiency of the verification activity. These could include use of checklists, checking and analysis using computer-based routines and computer-aided design (CAD) tools, templates, and other methods.

5.1 Design Reviews

Conducting technical reviews at critical points during the PCA design process greatly reduces the possibility of making poor or incorrect design decisions that eventually may result in the need to redesign the assembly. Although different companies' design processes may vary in detail, the sequence of activities to be accomplished to design a PCA is basically the same.

The main purpose of a PCA design review is to limit technical risk and assess design progress. The overriding objective of these reviews is to ensure that the PCA fulfills its intended purpose. An effective design review will

- Clarify technical requirements and evaluate the design for compliance with these requirements
- Challenge the basis for the design concepts
- Communicate requirements, design concepts, and detailed descriptions to other departments
- Determine what problems remain to be solved and what effort is needed to solve them
- Ensure that the PCA is testable, manufacturable, usable, safe, and reliable
- Provide additional knowledge and experience to the design team

The reviews are an important part of the PCA design process and are to be considered gateways that must be passed through successfully before proceeding to the next step in the process. Reviews may address either individual PCAs or groups of card designs if they are closely interrelated and the decisions that are made for one would probably affect the others.

5.1.1 Review Milestones in the Layout Process

Design reviews are most effective when they occur at the milestones indicated in the PCA layout process flowchart (Fig. 5.1). Regardless of where in the process reviews are held, they should all deal with the following general topic areas:

- Definition of requirements
- Evaluation of analysis and test data pertinent to the review
- Assessment of compliance of the design with the defined requirements
- Identification of potential problems and risks
- Identification of the potential impact of design decisions on "downstream" activities

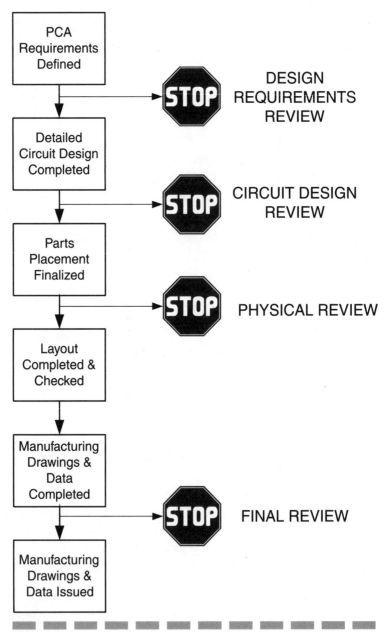

Figure 5.1
PCA layout process flowchart with key review points.

■ Determination if the design process should proceed to the next step(s) and identification of the risks associated with continuing

As shown in the flow diagram, there are four places in the process where formal design reviews should take place. These are

■ After the PCA's functional and physical requirements have been defined and documented and the basic approach for implementing the design has been established (design requirements review)

■ After the detailed circuit design has been completed (circuit design review)

■ After the parts placement has been finalized but before the interconnections have been routed (physical review)

■ After the design has been completed, and the manufacturing drawings and data are ready to be issued (final review)

These reviews should be a mandated part of the PCA layout process and considered gateways that must be passed through successfully before proceeding to the next step in the process (thus the use of the "stop signs" in the flow diagram). Although they may not be contractually required for a specific program, all these reviews should be made an integral part of the layout process.

The primary purpose for a *design requirements review* is to ensure that all requirements and constraints have been defined clearly and completely and are understood by everyone working on the project, and the potential is high that the selected design approach will satisfy all the stated conditions. Table 5.1 describes the activities that should be part of this review and the products or outputs resulting from each.

A *circuit design review* is held to verify that the detailed circuit design, when implemented in hardware and software, will result in a PCA that is likely to meet all its requirements. The activities that should be part of this review and their results are shown in Table 5.2.

The main objective of a *physical review* is to confirm, based on the placement of parts, that the product is manufacturable, testable, and maintainable within the constraints defined for the design. It

TABLE 5.1

Design Requirements Review Activities and Outputs

Tasks	Outputs
1. Identify key PCA functional/physical design requirements.	Documented key PCA design requirements
2. Describe recommended design/ manufacturing/test approach.	Documented PCA design description (functional block-flow diagram, electrical-mechanical density layout preliminary thermal-structural analysis)
3. Identify risk areas.	Documented risk areas
4. Evaluate potential impact of design approach on downstream activities/functions.	Documented PCA manufacturing approach
	Documented manufacturing test approach
	Documented functional test approach
5. Document action items resulting from the review and generate closure plan.	Documented review report and closure plan for open issues
	Decision to proceed with PCA design

also provides an opportunity to modify the placement prior to routing interconnections, since major movement of parts after routing is much more labor intensive. The activities that should be part of this review and their results are shown in Table 5.3.

A *final review* provides the last opportunity to assess the correctness and effectiveness of the PCA design before an investment is made in fabricating, assembling, and testing the end product. It also usually involves validation and formal approval (signoff) of the data and documentation that will be used to produce the assembly. The activities that should be part of this review and their results are shown in Table 5.4.

5.1.2   Preparation and Structure

Reviews should be initiated, organized, conducted, and documented by the person who has been assigned responsibility for

TABLE 5.2

Circuit Design Review Activities and Outputs

Tasks	Outputs
1. Review technical requirements.	—
2. Describe functional design approach.	Functional circuit design approach confirmed
3. Review schematic diagram and proposed PCA construction.	Detailed schematic design confirmed
	PCA construction confirmed (thermal/structural analysis)
4. Describe design risks and proposed mitigation plans.	Risk/manufacturing/test plans confirmed
5. Document closure plan for action items.	Review report (including action closure/risk mitigation/manufacturing/ test plans)
	Decision to proceed with PCA design

TABLE 5.3

Physical Review Activities and Outputs

Tasks	Outputs
1. Review results of action items from circuit design review.	Action item closure
2. Review HW design and layout rules and restrictions (program/PCA-specific requirements, routing requirements).	Finalized circuit design
3. Examine part placement layout and PWB construction.	Finalized part placement
4. Review PCA testability.	Validated PCA functional test procedure
5. Document/assign action items.	Action item list/closure plan
	Decision to proceed with PCA design

TABLE 5.4

Final Review Activities and Outputs

Task	Outputs
1. Review bare board manufacturing data and drawings for accuracy and completeness.	Validated and approved bare board manufacturing data and drawings
2. Review assembly data and drawings for accuracy and completeness.	Validated and approved assembly data and drawings
3. Review test documentation (data, programming and tools/fixtures) for accuracy and completeness.	Validated and approved test documentation

accomplishment of the PCA design. This review facilitator should issue a notice to everyone who will be participating in the review in advance of the meeting, and it should include an agenda and the names of the people who are to provide specific material at the review.

Participants in a PCA layout review should include representatives from project and functional organizations and key technical personnel involved in the design. Particular attention should be paid to involving people responsible for the overall system design, manufacturing, test, and integration, and especially the technical people concerned with the design of elements of the product that interface with the PCA being reviewed.

The reviewers should familiarize themselves with the PCA design requirements and other background information prior to the review and be prepared to discuss issues and concerns based on that information. A design review typically could be structured as follows:

1. Introduction (purpose, agenda)
2. Presentation of technical data
3. Discussions
   - Design requirements
   - Design approach
   - Requirements compliance
   - Risk issues

- Design verification
- Effects on other activities
- Modifications to the design
- Actions to be undertaken and a plan for performing them
- Next activities to be accomplished

Other, nontechnical design constraints and issues may be considered during the review if they are identified as being appropriate. These could include

- The design schedule
- Budget status, present and projected
- The estimated cost of the finished PCA

5.1.3   Outputs from the Review

Reviews should be considered successfully completed when the following results have been achieved:

1. All required inputs to the review were provided and evaluated.

2. The impact of the design approach on downstream activities was assessed.

3. All new activities identified and assigned during the review have been documented, along with a closure plan indicating responsibility assignments and closure dates.

4. There is agreement on proceeding to the next step(s) in the design process, identification of the risks in taking the agreed action, and how these risks will be managed.

5. The review facilitator has issued a review report containing copies of inputs to the review, all data presented at the review, the assigned action closure plan, and a summary of decisions and recommendations made.

Standard verification procedures can be defined and documented by large and small design organizations and are equally valuable to both. Large organizations tend to prepare formal and structured documentation and exercise rigorous control over changes. Small companies do not necessarily need such an investment. Verification procedures, methods, checklists, and other tools can be documented informally and easily distributed and controlled within small design organizations. This is a relatively minimal investment that is far outweighed by its downstream value. Avoidance of a few redesign cycles usually will more than repay the cost.

5.2 Checking

Quality inspections should be performed on documents, data, and other outputs during the PCA design activity. These checks are key to producing high-quality results. In addition, capturing, collating, and assessing the results of checking activities over a period provides valuable information for improvement of both the design and checking processes.

It is important that standard procedures and tools or aids be established for performing checking activities. This promotes predictability and repeatability of results and reduces the opportunity for error caused by inconsistent application of a checking methodology.

Exactly where in the design process checking should be implemented and the methods to be applied are greatly dependent on the tools and techniques used to produce the design. The checking approach for PCAs designed and documented using manual techniques is quite different from that used if the design is produced using computer-based tools. Manually produced designs usually require more checkpoints and more time expended per checkpoint. Manual checking also is more error-prone, especially for complex PCAs.

5.2.1 Basic Checkpoints in the Design Cycle

In general, the integrity of a PCA design should be confirmed by checking in the following primary areas as the process proceeds:

- The mechanical definition
- The schematic data

- The physical and functional parts library
- The completed layout
- The resulting drawings and manufacturing data

5.2.2 Checking Methods

Within each area, the method of checking will vary depending mainly on how the material was created.

CHECKING MECHANICAL DATA A CAD system may be used to create a PCA's mechanical description, including definition of the card outline, tooling holes, dimensional origin, machining requirements, physical part locations, connector placement, keep-out areas, and other mechanical features. If these data can be input directly into the CAD system that will be used to lay out the board, little or no manual checking is required to validate the data. As long as there is no need for human or machine intervention to modify or adjust any of the data, checking at this point can be limited to verifying that the transfer of data was completed successfully and no anomalies (data error messages) were generated. The designer should be aware that if the data must be converted, translated, or re-formatted before or during the transfer, errors may be introduced that must be found and corrected before the data are used.

If the mechanical description must be transcribed manually from drawings, then the resulting data should be checked carefully before they become the basis for further design activities. Employment of an incorrect mechanical dimension ultimately may result in production of hardware that cannot be used as built and must be either modified or, worst, scrapped and the assembly redesigned. The primary purpose of the checking done here is to confirm that the mechanical requirements have been transferred accurately to the design environment.

CHECKING THE SCHEMATIC DATA The same approach is applicable to input of circuit (schematic) data. If part and interconnection data can be transferred directly between systems without change and no anomaly reports are generated, manual checking is not required to validate the data as a good input to the layout environment. This statement does not imply that the circuit will function as required, only that the schematic data have been captured accurately. It is always incumbent on the circuit designer to verify the functionality of the design using modeling, simulation, breadboarding, testing, or any other appropriate method. There is, however, the issue of whether the schematic is consistent with any predefined design rules, and this can only be verified by reviewing the schematic. Some computer-aided engineering (CAE) systems used for circuit design allow the user to program in the design rules for a project and have the system check for conformance to those rules.

Manually created schematics do not provide the advantage of automated input to the design environment. If a CAD system is used for PCA layout, then a labor-intensive effort is required to convert the information on the diagram into data that the CAD system can recognize and use and manually input them into the design system. This conversion and input effort is potentially a significant source of error, and the resulting data must be checked carefully, also done manually, before they are used. Printed circuit CAD design systems usually include capabilities for some checking of the validity of circuit inputs, but they are relatively nonintelligent and usually will catch only format-related errors and illogical statements.

For a manually laid out PCA, there is no need for data conversion and input, but when the design is complete, a detailed check must be made to verify that all interconnections were made correctly. This is also a manual, labor-intensive activity.

CHECKING THE PARTS LIBRARY DATA   The primary source of data for a PCA design resides in a functional and physical parts library, which provides information about the electronic components (and sometimes mechanical parts) that are to be used in the assembly. This is the case whether an assembly is designed using computer-automated or manual methods. Physical parts libraries contain information that defines the shape of parts to be used in a PCA. This includes body dimensions, lead locations and identifiers, footprint data (pads, leadholes, vias, etc.), reference designator type and location, and a reference or linkage to functional and schematic information about the parts that may be located in another library or another part of the same library.

Functional libraries usually define performance-related parameters of parts. They include schematic symbology, design rules for part utilization, parametric data used for circuit simulation and modeling, thermal characteristics, and links to physical part information.

If computer-aided systems are used to develop and document a circuit schematic and lay out a PCA, then the library data will be used directly by these systems as the basic information required to represent the physical and functional parts included in the design. For manually prepared schematics and assembly layouts, the part information is usually captured from the library and used to manually create circuit symbols for the schematic and "doll cutouts" or templates that represent the physical parts for the board layout. The library source in this case may be nothing more complicated than an organized collection of vendor data and predrawn dolls.

The information required for parts to be inserted into these libraries is usually obtained from a manufacturer's technical specification or data sheet that describes the part's functional and physical characteristics. This data must be abstracted manually from the specification and inserted into the appropriate library. Computer-aided system libraries usually require that the data be input in a specific, complex structure and format. It is necessary that every part library input be checked thoroughly to verify that

■ The data have been transcribed accurately and completely.

■ It is structured correctly.

■ The manufacturer's information has been verified and is correct and up to date.

■ Any possible ambiguities in the data have been resolved.

CHECKING THE LAYOUT   The completed PCA layout must be checked before preparation of drawings, artwork, and manufacturing data is begun. The primary objective of this check is to verify that the design database completely and accurately represents the intended design in terms of electrical continuity, conductor dimensions and spacing, part spacing, orientation and location, use of the correct parts, and application of the appropriate circuit and physical design rules.

Automated design systems have the capability to perform many of the checks required to verify the correctness of the layout. The CAD software program can be directed to verify that the interconnections are consistent with the schematic input, that no conductor width and spacing rules have been violated, that the correct parts have been used, and that none of the general design rules have been violated. Most CAD programs are designed to perform these checks "online" and in the background in order to prevent errors while the design is being created. Unfortunately, some CAD systems allow these built-in checks to be disabled to provide design flexibility. This defeats the integrity of the system's checking capability and mandates that a checker independently verify that no design requirements, as defined by the original data inputs, have been violated. Usually, a separate verification of the design database can be performed on the system to confirm this.

For manual layouts, each design feature must be checked, because they are all a potential source of error. This should include, at minimum, actual interconnections compared with the schematic diagram and use of the correct part types and part dimensions. Also, line widths, and spacings, part sizes, hole sizes, land-to-hole ratios, edge-spacing and keep-out violations, and dimensions and locations of board features (mounting holes, tooling holes, attachment of mechanical parts, etc.) should be verified.

CHECKING THE DRAWINGS AND MANUFACTURING DATA   A PCA design is turned into a product through the creation of a set of drawings and data that are used to manufacture and test the product. The extent of the ability of computer-based systems to generate the required documentation varies significantly. Direct output of drawings based on information taken from the design database minimizes but does not completely eliminate the need for detailed checking. If design data must be transferred from a PCA design system to a mechanical drafting system to generate manufacturing drawings, than the result should be examined more closely. This is to ensure that no errors were introduced due to human or machine intervention.

PCA design systems also produce outputs in the form of data files that can be used to perform a variety of manufacturing and test functions. The primary output is data used to produce a variety of phototools needed to fabricate the bare circuit board. If the phototools are produced by the design function, then they must be checked before being used for manufacturing the boards. If the correctness of the layout has been confirmed, then the data used to drive the photoplotter are also correct. It is mainly the quality of the films that the plotter produces that should be checked. Photoplotting is a complex process requiring a high degree of precision, and there is no guarantee that artwork of acceptable quality will be produced consistently. Many opportunities for error exist, and close examination of the artwork should be performed. In addition, blemishes can occur easily that if not identified could result in fabrication of nonfunctional (scrap) boards. Typical film blemishes include pinholes, scratches, and emulsion defects.

Other system outputs used for manufacturing and test include data that can be used to drive numerical controlled (NC) fabrication equipment (drills, board profilers, etc.), assembly machines (pick-and-place, adhesive dispensers, etc.), and automated test equipment. Since these are also derived from the design database, the only checking usually required is to ensure that the files have been produced in the correct format and that they are derived from the correct design.

If manufacturing drawings and phototools are prepared manually, then a much more extensive checking effort usually is needed to verify that all data produced accurately portrays the design. For artwork, the master should be checked to ensure that all interconnections have been transposed accurately from the layout. The dimensions and placement of conductor features must be measured, especially land areas, line widths, and spacings. For multilayer boards, layer-to-layer alignment of pads for through-holes and vias should be checked carefully. Drawings should be checked for correctness of pictorial and dimensional data, instructional notes, and parts lists.

5.2.3   Checking Tools

The features of a PCA design produced with a CAD system are controlled by the data and design rules established in that system.

Outputs of the system, such as reports, listings, and hard-copy plots, are the primary tools used to check the design. Net and parts lists can be used to verify that the schematic input was created correctly. Drill and pad lists provide assurance that these features are placed correctly and are sized properly. Design rule verification listings ensure that there are no physical or circuit violations in the design. Check plots can be produced to show part placement, conductor routing, and drilled hole locations. The basic mechanical features of a board also can be plotted, detailing outline, tooling, and mounting-hole dimensions. Part descriptions resident in the library can be listed and plotted separately to check the accuracy of the data against the latest vendor information.

The quality and accuracy of photoplotted artwork usually are checked using optical inspection and measuring equipment. This can be done manually, a labor-intensive activity that potentially introduces a significant opportunity for error due to missed defects and lack of consistency. Use of automated optical inspection (AOI) equipment, when properly programmed, can locate and identify a very high percentage of artwork defects, especially for dense circuit boards. Although AOI is usually used to verify the quality of fabricated circuit layers prior to multilayer board lamination, the technology has become recognized as an excellent method for checking both master and working artwork.

Checking manually laid out PCAs involves the use of a different set of tools, especially if the artwork masters and manufacturing drawings are also created by hand. The layout must be checked against the schematic systematically, point by point, and usually involves lining out each connection on both the schematic diagram and a copy of the layout. For multilayer designs, this is a tedious, time-consuming task. Layer-to-layer alignment of tooling and mounting holes, pad stacks, and other features must be checked. The artwork and drawings are produced using the layout as the information source. These also must be checked carefully to verify that the data were transcribed correctly from the layout. This is especially true for manually prepared artwork, since every trace, pad, and feature must be placed by hand.

5.2.4 Checking Documentation

The results of checking activities should be recorded and the data retained for each PCA. If design-related problems arise during fabrication, assembly, test, or integration, review of checking documentation is usually a key to determining the source of the problem. In addition, checking data provides valuable information for design and checking process improvement. The kinds of discrepancies identified during checking can help establish which defect types are most common so that improvements can be focused in those areas.

If checklists are used as an aid to ensure that all required verification activities were performed, these should be part of the retained data. After completion of a design, all checkplots, marked prints, listings, and other checking-related documentation for a PCA should be gathered together into a file and stored for easy retrieval. Copies of meeting notes, review reports, and design analyses also should be included. Significant time can be saved by being able to quickly retrieve and review all technical data for a PCA not only for problem solving but also if design changes are required in the future or the assembly can be used for a different application.

5.3 Producibility Evaluation

A major part of the designer's responsibility is to ensure that a PCA can be manufactured in a cost-effective way. The physical design of a PCA has a direct effect on its producibility because it is the major factor in determining which process can be used to build it. Key features of a design that most directly affect its manufacturability are

- Part type selection and part placement and orientation
- Printed circuit board material and board construction
- Conductor geometry and dimensions
- Overall complexity and circuit density of the PCA
- Type of circuitry and criticality of circuit performance

Producibility should not be an afterthought, to be considered only after a layout has been completed. It needs to have equal status with other design requirements and constraints. Unfortunately, absolute rules about cost-effective design are extremely difficult to establish due to the complexity of the manufacturing process and the broad variability of elements found in PCAs. Because different types of designs have varying cost sensitivities, easy-to-use, globally applicable cost models are hard to find.

There are, however, some basic relationships that can be used to make fundamental design decisions. These will be discussed in the following sections of this chapter. In addition, cost models have become available that can be tailored to accommodate the unique types of assemblies built by a company and take into account its unique manufacturing and assembly processes.

5.3.1 Part Type Selection, Density, and Placement

How easily a PCA can be manufactured generally depends on the types and quantities of components used and how they are mounted to the circuit board. PCAs usually are classified for manufacturing as using plated through-hole (PTH) devices, surface-mount technology (SMT), or mixed technology (through-hole and surface-mount). For ease of manufacturing, a PCA is preferred to be all single-sided PTH or SMT to minimize the number of different assembly and soldering operations. If mixed technology is required, all parts should be mounted on the same side of a board. Mounting parts on both sides significantly increases manufacturing difficulty and should be used only when dictated by circuit density or performance requirements. Table 5.5 is a generalized listing of assembly types in order of increasing cost to manufacture.

Most PCA designers and engineers are familiar with the manufacturing process for PTH assemblies. However, the manufacturing flow of an SMT or mixed-technology assembly is much more complex. It depends on many different design features, including the number, type, and location of individual SMT devices (and PTH components), the manufacturing and test equipment to be used, and the process experience of the manu-

TABLE 5.5

Listing of Common Assembly Types in Order of Increasing Cost to
Manufacture

Type of Assembly	Relative Cost
Through-hole components mounted on a single side	Least costly
Surface-mounted components on a single side	
Surface-mounted components on both sides	
Mixed technology (through-hole and surface-mounted components) on the same side	
Mixed technology with through-hole components on one side and surface-mounted components on the other side	
Mixed technology with through-hole components on one side and surface-mounted components on both sides	Most costly

facturer. It is therefore important that the manufacturing and
test engineering organizations be consulted early in the design
cycle to discuss and review process issues, component selection
guidelines, and cost tradeoffs.

Following the manufacturing preference guideline in Table 5.5
should minimize the PCA manufacturing cost and cycle time.
It should be noted, however, that a product's total life-cycle cost is
to be considered when evaluating design options. This involves
weighing nonrecurring design costs against recurring manufacturing costs, as well as considering trading component costs
against their availability, reliability, maintainability, reparability,
and performance.

GENERAL PART SELECTION GUIDELINES

▪ Choose parts having the greatest possible lead spacing.
Manufacturing yield loss is a function of part lead spacing
and rises exponentially as spacing decreases. Higher lead
density requires the use of special manufacturing processes to
minimize fabrication, assembly, and soldering problems, which
slows down production.

- Minimize the number of different part types and part numbers used on an assembly. Select as many parts as possible from a preferred or standard parts list, if one exists.

- Avoid parts having electrostatic discharge (ESD) sensitivity under 100 V. Those sensitive to less than 30 V should not be used unless absolutely necessary, because they may require use of special handling processes and materials.

- Minimize use of parts requiring heat sinks, thermal pads, thermal grease, or thermal adhesives. Their effectiveness to remove heat from a component usually depends on an operator's ability to apply them consistently and uniformly. The additional material and manufacturing labor costs justify performance of a thermal design analysis to determine if supplementary heat-removal elements are required.

- Parts whose weight exceeds 3.5 g per lead may have to be secured to the circuit board using mechanical brackets, clamps, or adhesive. The additional material and manufacturing labor costs should be considered when selecting part types.

- Radial leaded components should be specified with leads that meet a PCA's lead length requirement to avoid expensive trimming operations. This is especially critical for stud-mounted parts and parts with multiple hard pins, such as PGAs and connectors. Select parts with built-in spacers or standoffs whenever possible to facilitate soldering and cleaning under the parts. If separate spacers are required, take care that they do not interfere with soldering or cleaning operations.

- Axial leaded components that are designed to be mounted vertically should be avoided because they may require hand installation and additional mechanical support.

- Select parts that are compatible with the materials that are to be used to manufacture the PCA, such as fluxes, solvents, and cleaners. Avoid parts that use plastics that can be degraded by processing temperatures (such as baking, cleaning, or soldering) and which have internal construction using metallic interconnection materials (such as solder) that have a melting point close to or lower than that of eutectic solder.

■ All parts to be soldered should be specified with pretinned terminations. Gold-coated terminations should not be soldered because the resulting gold-tin intermetallic is brittle and could cause premature solder joint failure. For parts to be used in plug-in applications, a hard, corrosion-resistant material (preferably gold) should be specified.

■ Avoid connectors that straddle a board because they are difficult to mount and may require a special soldering operation. Avoid connectors that require each termination to be soldered individually. Avoid connectors using insulator material that may be affected by manufacturing processes or have a nonuniform body cross section, which may cause warpage during soldering.

■ Minimize component preparation activities by specifying parts that are ready to be assembled (proper lead lengths, terminal finishes).

■ Avoid parts requiring manual assembly and soldering, such as ones that are odd sized or odd shaped, are unusually oriented, or cannot withstand mass soldering processes. Hand soldering is a relatively uncontrolled process, and the quality of the end result is highly dependent on an operator's skill and training.

SMT PART SELECTION GUIDELINES

■ When choosing SMT parts, it is critical that standard body sizes be used (preferably parts that conform to JEDEC specifications). Special or unique parts can cause major manufacturing problems, since an odd body size or shape may not always be programmable into existing pick-and-place equipment, which means they must be hand placed and possibly damaged due to additional handling.

■ Use rectangular rather than square parts whenever possible. Square parts, especially small ones, are difficult to handle, and their orientation may not be reliably identified by automated placement equipment. In addition, their shape and low mass make them prone to tombstoning (standing on end) during soldering. Avoid parts less than 0.100 in long, especially if they are to be mounted on the solder side of a PCA, where presolder adhesive may obscure the circuit board pads.

▪ Avoid parts that have porous, convex, or concave surfaces because they are difficult to handle with automatic pick-and-place equipment. Smooth, flat-surfaced parts are preferred. Avoid parts with heights greater than length or width. Parts with these aspect ratios are unstable during handling and subsequent soldering operations.

PART PLACEMENT GUIDELINES

▪ Be aware of and adhere to part spacing rules. Clearances between parts dictate the percentage of parts that can be placed on an assembly using more reliable, less costly automated processes.

▪ Place all parts of the same type in the same orientation for ease of assembly. This is especially important for polarized and multipin parts.

▪ Determine what soldering process is planned to be used in production and how parts should be placed to maximize solder joint quality. If an assembly is to be wave soldered, have manufacturing identify the assembly's probable orientation through the wave, and solicit recommendations for preferred part orientation and spacing. Parts should be placed and spaced to minimize the possibility of shadowing their respective terminations, resulting in solder-starved joints that require subsequent manual touchup.

▪ Avoid clustering of parts, where some areas of an assembly are densely packed and other areas are relatively empty. A balanced layout allows use of less costly manufacturing processes (board fabrication, assembly, and test) and usually makes the design much easier to implement.

5.3.2 Board Material

The raw laminate material used is one of the largest cost components in a fabricated circuit board. When selecting board material (laminate and prepreg), the designer should be aware of the cost impact of

- The type of dielectric material specified (FR4, polyimide, etc.)
- The type of reinforcement specified (standard glass, high-T_g glass, woven, random fiber, paper, etc.)
- The electrical (dielectric constant, dissipation factor, etc.), physical (peel strength, moisture absorption, flammability, etc.), and thermal [glass transition temperature, coefficient of thermal expansion (CTE), etc.] properties specified
- The type, thickness, and combinations of copper foil required
- The material thickness and tolerance specified
- The material specifications and certification specified
- Availability from material fabricators

It is also important for the PCA designer to find out which suppliers may be fabricating the bare board. This information could enable the designer to identify what materials the potential supplier(s) prefer because of existing inventory and processing experience and capabilities. The cost and quality of the finished product may benefit if it is possible to consider fabricator preferences when specifying board materials.

5.3.3 Board Construction and Dimensions

PANELIZATION Circuit boards typically are processed in multiples on a single panel. Since board material is a major cost factor (up to 30 percent of the total), efficient panel use is critical. Designers should be aware of the circuit board fabricator's "standard" panel sizes and manufacturing capabilities. Use of larger panels is usually more cost-effective for a fabricator, resulting in more efficient processing, better equipment use, and reduced handling/racking. Smaller panels, however, provide the greater dimensional stability necessary for successful fabrication of dense designs. When establishing the outline dimensions of a circuit board, a designer should determine what the board fabricator needs for panelization borders, pattern spacing, tooling holes, plating thieves, fiducials, etc.

Figure 5.2 shows four individual circuit boards on one process panel and indicates some typical dimensions for keep-out zones at the edges of the panel and minimum distance between the circuit patterns. Generally, board cost is the cost for processing the panel divided by the number of individual boards on the panel. Panel utilization efficiency is the total area of the boards on a panel divided by the total panel area. Panel utilization of 85 percent or greater is very good, whereas less than 65 percent is considered by manufacturers to be poor material utilization efficiency. Table 5.6 identifies standard process panel sizes used by many industry fabrication shops.

TOOLING HOLES Tooling holes act as alignment guides for fabrication, assembly, and test fixtures. Tooling hole patterns for each family of boards (i.e., boards of the same size and shape) should be the same to minimize the variety of tooling required. These holes also should be used to establish the PCB datum, and all board dimensions and component locations should be refer-

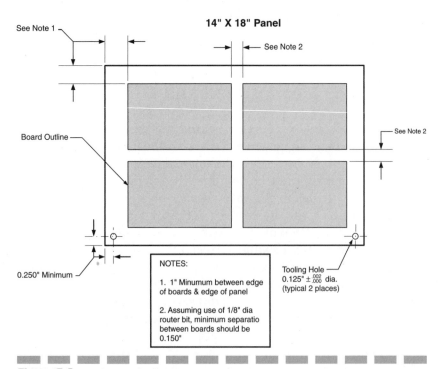

Figure 5.2
Panel layout showing typical dimensions for spacing the circuit boards.

TABLE 5.6

Commonly Available Process Panel Sizes and Allowable
Circuit Board Dimensions

Standard Panel Size (in)	Largest Circuit Board (in)
16×18	13×15
16×28	13×25
16×36	13×33
18×21	15×18
18×24	15×21
18×41	15×39
21×24	18×21
24×28	21×25
24×31	21×28
24×36	21×33
24×42	21×39
30×36	27×33
30×41	27×38
30×44	27×41
36×42	33×39

Note: Circuit board dimensions may be increased somewhat, depending on its thickness. The board fabricator should be consulted for this.

enced from this datum. They are inside the board outline and
are independent of panel tooling holes that are provided outside
the outline by the board manufacturer.

Three tooling holes, located at the corners of the board at $(0,0)$,
$(X,0)$, and $(0,Y)$, are preferred. At minimum, boards should contain two tooling holes located on a diagonal and spaced as far
apart as possible. Standard tooling holes should be nonplated,
having a diameter and tolerance consistent with the board fabricator's requirements. A typical tooling hole dimension might be

0.125 in diameter with a tolerance of $+0.003/-0.000$ in. Also, component land patterns, probe points, and fiducials should not be located near the tooling holes to facilitate fixturing. A minimum clearance of at least 0.075 in should be maintained around the tooling holes.

FIDUCIALS The use of fiducials facilitates PCA manufacturing by allowing use of automated equipment equipped with machine vision systems. Fiducials are normally used for SMT or mixed-technology designs. PTH assemblies do not require fiducials. A typical fiducial shape is a circle of 0.050 in diameter minimum, as shown previously in Fig. 4.6. There are three different types of fiducials: component, board, and panel-array. Component fiducials are frequently desired by the manufacturer to allow accurate placement of individual large, leaded SMT components, fine-pitch SMT devices, and ball-grid array (BGA) devices. Two component fiducials usually are needed per device. The designer should consult with the manufacturer to identify the types of parts needing fiducials and where they should be located with respect to the parts.

Board fiducials allow a vision system to make adjustments in the X, Y, and theta (rotational) directions for general fabrication and assembly activities. Each circuit board should have fiducials located at three corners on both external surfaces. At a minimum, two fiducials located as far away from each other as possible on the longest board diagonal should be provided per side. Panel-array fiducials are used when an array of identical boards are fabricated and assembled before they are separated from the panel. They serve the same basic function as the board fiducials.

BOARD CONSTRUCTION Once a designer has decided how many layers are needed for a design, the method of construction (stackup) should be determined. It is very important to obtain recommendations from the intended board fabricator in order to establish a cost-effective construction that can be built using off-the-shelf materials and standard processes. In addition to specifying the number of layers required, the position of signal and power/ground layers in the stackup, the copper weight to be used for each layer, the basic laminate type, and layer-to-layer spacing should be identified. Although board properties generally are based on electrical, thermal, and physical requirements, overspecifying every construction detail tends to constrain the fabricator (many times unnecessarily) by limiting its ability to use materials and processes that could provide better yields at lower cost. Some of the construction features that could add significantly to the cost of a board are

- Overall board thickness greater than 0.175 in or less than 0.028 in
- Board thickness tolerance less than ±8 percent (±10 percent preferred)
- Unbalanced construction (dielectric, copper, layers)
- Laminate core thickness less than 0.005 in
- Copper weight greater than 1 oz ($^1/_2$ oz preferred)
- Mixed copper weights on core material greater than 2:1 (1:1 preferred)
- Uneven conductor distribution on a layer

BOARD DIMENSIONS   The accuracy of board outline dimensions and the locations of machined, drilled, and hardware installation features should be reasonably specified. Unnecessarily tight tolerances potentially can cause a manufacturer to scrap otherwise good-quality products, and this will be reflected in a greater projected cost. The following are examples of preferred specifications:

- Positional tolerance of a dimension of a routed or machined feature should be no less than ±0.005 in.
- Spacing between inner or outer layer conductor and the edge of the board or machined feature should be greater than 0.010 in.
- Internal radii should be no less than 0.035 in.
- The tolerance between a drilled datum hole and the board edge and edge-to-edge dimensions should be no less than ±0.005 in.
- Unplated slots should be no less that 0.070 in wide, with dimensional tolerances no less than ±0.005 in.
- The preferred router diameter is 0.093 in, with inside routed corners of 0.047 in radii or greater.
- An edge bevel angle of either 30 or 45° is preferred.

CONDUCTOR, PAD, AND HOLE DIMENSIONS Preferred dimensions for inner and outer layer conductors, spacing, annular ring, plating thickness, and tolerances are shown in Table 5.7. The preferred dimensional relationships between holes and pads on inner and outer layers are shown in Fig. 5.3.

After material, the expense of drilling is the second largest cost element of a circuit board. Very small or nonstandard diameter holes having tight diametral and positional tolerances are major contributors to this cost, as are designs that use minimal annular rings. Preferred guidelines for drilled holes include the following:

- Drill sizes should be 0.0135 in or greater.
- Annular rings should be no less than 0.005 in when tangency of the hole to the edge of the pad is not allowed.
- Positional tolerance should be no less than 0.004 in radial true position.

TABLE 5.7

Preferred Dimensions for Circuit Board Features

Feature	Preferred Dimension (All Linear Values in Mils)
Outer layer traces and spaces	4–5
Inner layer traces and spaces:	
1 oz copper	3.25–3.75
$^1/_2$ oz copper	2.5–3
Hole diameter, through vias	8–10
Hole diameter, blind or buried microvias	3–4
Minimal annular ring	5–10 over finished hole
Plated hole diameter tolerance	±2–3
Copper plating	$^1/_2$–1 oz.
Solder plating	1

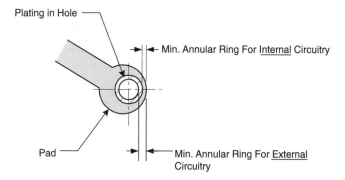

Minumum Pad Size Should = Max. dia. of Finished Hole + 2x Min. Annular Ring (Table 1) + Manufacturing Allowance (Table 2)

Table 1 Minimum Annular Ring

TYPE	COMMERCIAL/ INDUSTRIAL	MILITARY/ SPACE
External-Supported	.002"	.004"
Internal-Supported	.001"	.002"
Unsupported	.006"	.010"

(NOTE: Min. annular ring shall not be less than max. allowable etchback)

Table 2 Manufacturing Allowance

GREATEST BOARD DIMENSION	ALLOWANCE (Incr. Complexity ←)
Up to 12"	.010" ← .020"
Up to 18"	.018" ← .024"
Up to 24"	.022" ← .028"

Figure 5.3
Dimensional relationships between holes and pads.

■ Diametral tolerances for unplated holes should be no less than ±0.002 in and for plated holes should be no less than ±0.003 in.

ELECTRICAL CHARACTERISTICS Many of today's PCAs contain complex, high-speed, high-density circuitry. For this type of design, the circuit board serves as more than just a means of interconnecting parts; it becomes a functional part of the circuitry. It therefore becomes necessary to specify that the circuit board, and in some cases the finished assembly, meet precise electrical characteristic requirements.

Electrical properties of laminate material, such as dielectric constant and dissipation factor, are determined by the type of material specified, as is the bulk resistance of the conductor material. These parameters are standard throughout the industry, and laminate manufacturers exercise reasonably close control over their consistency. Layout and location of the conductors (length, width, spacing, parallel runs, stubs, interlayer relationships, etc.) also have a significant effect on circuit characteristics.

For high-speed circuitry (fast rise/fall times for digital circuitry, high frequency for analog circuitry), the characteristic impedance of the conductors on a board is extremely important. If this parameter is not well controlled on a board and impedance mismatches occur in the circuitry, PCA performance can be severely affected, resulting in signal degradation, timing problems, and false triggering. For critical circuitry, characteristic impedance should be required to be kept within 10 percent of a specified value (e.g., between 45 and 55 Ω for a nominal value of 50 Ω).

Controlled-impedance circuit boards should be required only when absolutely necessary, since they must be fabricated to very exacting specifications. Tolerances on conductor dimensions and spacing must be held more closely than for standard boards. Electrical and physical properties of the board materials (conductor, laminate, and B-stage) must be tightly controlled. Finally, requirements for specific layer-to-layer spacing between conductors and thickness of dielectric material for multilayer boards put a heavy burden on the fabrication process. In general, controlled-impedance boards are much more expensive than standard boards because they require additional labor and more expensive materials and have lower yields.

5.4 Testability Evaluation

Testability is a measure of the assurance that can be provided by tests that a PCA will perform its intended functions properly. Three common types of tests are most frequently used to accomplish this:

- *Bare board testing,* which checks for manufacturing defects

- *In-circuit testing,* which confirms that after assembly, all parts function and are interconnected properly

- *Functional testing,* which verifies that the PCA, when integrated with the rest of the system, will perform as required

In addition, when a severe operational environment is involved, the PCA may undergo environmental testing to ensure that it will function properly. For certain types of systems, self-testing capabilities may be built into the PCA itself in order to monitor its performance during system operation.

Demands on today's sophisticated PCAs make it critical for the designer to be directly involved in determining how the product will be tested and accommodating the resulting test requirements and constraints during the design activity. Testability of the product cannot be considered as an afterthought. Tight time-to-market constraints are being driven by global competition in the electronics industry, and design iterations that used to be necessary to accommodate test requirements are no longer tolerated.

High-density circuit boards with closely spaced, fine-pitch components make it difficult to impossible to add test capabilities after a layout has been completed. Unless a layout has incorporated provisions for test, it probably will require significant modifications or possibly need to be completely redesigned.

5.4.1   Bare Board Testing

It is very important that bare complex circuit boards be tested for product quality prior to assembling large quantities of expensive components on them. Multilayer boards with fine lines and spacings, high plated hole aspect ratios, blind or buried vias, and controlled electrical parameters tend to have lower than normal yields due to the inherent difficulty in processing them. Finding shorts or opens on inner layers or noncompliance to parametric requirements after assembly may have a major impact on a program.

A bare board's electrical and mechanical characteristics usually are verified by the fabricator using test coupons that are defined by and included on the circuit artwork. It is important that the designer interface with the board fabricator to determine the best configuration to use for the test coupons and where they should be located.

Checking for conductor shorts and opens on layers and/or a finished board usually involves using probe-type test fixtures. On complex, high-density multilayer boards, the designer must make sure that the ends of every conductive feature are accessible for contact by the tester. In some cases this may require addition of probe contact points to the circuit layout.

5.4.2 In-Circuit Testing

Present-day in-circuit testing facilities provide the ability to verify both component functionality and manufacturing flaws after parts have been assembled on a board. The capabilities of in-circuit testers have expanded to the point where they also can also confirm the performance of some multipart circuit functions.

In-circuit testing usually is applied in a production environment to verify that the parts are connected to each other correctly and that the parts themselves are operating within their specified parameters. It also can be used to determine if the assembly process is under control and assess the quality of the parts being used by analyzing test data gathered over a specific period of time. In most instances, specialized fixtures containing spring-loaded probes that make contact with each circuit net or component pin within a net are used to perform in-circuit testing. Figure 5.4 shows an example of a PCA test fixture that interfaces with the in-circuit test equipment. It is preferable that all probe contacts be made on one side of a PCA, since a double-sided fixture is much more complex and costly than a single-side fixture.

Requirements for in-circuit testing should be defined early in the PCA design process and should include such considerations as

- How unused device pins are terminated
- Provision of pull-up and pull-down resisters
- Probing of unused active pins
- Measurement of specific device parameters

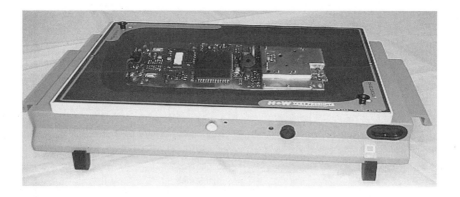

Figure 5.4
An example of a PCA test fixture that interfaces with automatic test equipment.
(Used by permission of H + W Test Products, Inc.)

- Measurement of specific net characteristics
- Method of measuring power and ground
- Exercising test capabilities built into devices, such as boundary scan
- Physical placement and distribution of test point locations
- Test point configuration

Figure 5.5 contains examples of some typical test point configuration and location constraints. Specific requirements, which are determined by the type of test equipment and fixtures being used and the physical configuration of the PCA, should be obtained from the organization that will be performing the in-circuit tests. The designer must define the location and designation of each test point on a PCA and identify what net and/or device it is associated with. These data are needed for fabricating the pin fixtures and for programming the tester. Also, holes for aligning the test fixture with the assembly must be provided in the circuit board.

TESTING DIGITAL CIRCUITRY   In an unpowered mode, circuit opens and shorts can be identified at preprogrammed thresholds. In a powered mode, test patterns drawn from a

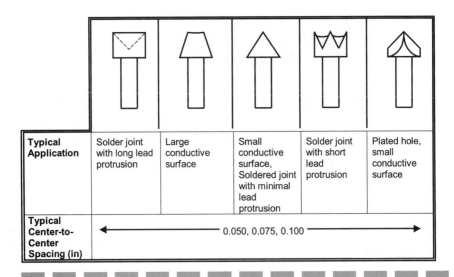

Typical Application	Solder joint with long lead protrusion	Large conductive surface	Small conductive surface, Soldered joint with minimal lead protrusion	Solder joint with short lead protrusion	Plated hole, small conductive surface
Typical Center-to-Center Spacing (in)	← 0.050, 0.075, 0.100 →				

Figure 5.5
Typical test point configurations, utilization, and spacing.

library of patterns can be injected into digital components using a broad range of clock speeds.

Self-testing capabilities built into parts can be exercised, and artificial circuit faults can be injected into the circuitry to check its ability to recognize them. Flash memory can be programmed and checked, and the overall testability of a circuit can be analyzed.

TESTING ANALOG CIRCUITRY Circuit resistance, capacitance, and inductance can be measured, and polarized SMT and leaded capacitors can be checked for proper installation. Parametric tests can be performed on diodes, transistors, FETs, Zeners, and other analog devices. Operation of fuses, switches, potentiometers, and connectors also can be tested. Functional testing can be performed on operational amplifiers, comparators, voltage regulators, and other analog circuitry.

IN-CIRCUIT TEST EQUIPMENT In-circuit test equipment is available with a broad spectrum of capabilities, ranging from simple machines that check for circuit opens and shorts, to testers that measure device parameters, and on to systems that dynamically exercise device and circuit functions. All systems need to be programmed to perform their intended test functions.

To perform functional testing, device models must be input into the system to automatically generate test patterns and accomplish circuit analysis. For each device type, the function of each pin must be identified along with its response to a variety of signal patterns. Once a model is entered in the test system's library, it can be used to confirm the functional operation of physical parts on a PCA.

5.4.3 Functional Testing

Limited functional testing can be performed by in-circuit testers, but completely exercising all circuit capabilities on a PCA usually requires the use of a dedicated assembly-level functional tester. The basic objective of PCA-level functional testing is to simulate an assembly's operation as if it were installed in the final product. This type of testing is usually implemented for large-volume, high-rate production or for very complex systems. In both environments, locating and replacing an out-of-spec PCA, once it is installed, can be a very time-consuming and disruptive operation. For PCAs that have variable components that must be adjusted to meet circuit performance requirements, the functional tester can be used to implement this activity.

To perform functional testing, the PCA's input-output (I/O) connectors (in most cases, edge connectors) are attached, usually by a set of cables, to the tester that then inputs a set of signals to the card and measures its response. It is important that the PCA designer be aware of the type of testing required for an assembly and the physical configuration of the test equipment to be used and accommodate these requirements in the layout by providing proper access to all connectors and variable components.

5.4.4 Environmental Testing

Environmental tests are used to reveal how products will stand up to the damaging effects of a variety and combination of elements.

They are designed to simulate or exaggerate the conditions under which a PCA will be required to operate. These conditions can include temperature extremes, altitude, high humidity, rainfall, salt spray, vibration, and shock (both thermal and mechanical).

A PCA may be tested environmentally for a variety of reasons: to verify the design approach, to predict its long-term reliability, or to ensure that it is being built properly. The assembly may be tested in conjunction with the final product or as a stand-alone item. For design verification or reliability prediction, the product is usually exposed to a combination of the harshest conditions it is expected to experience. If a PCA can withstand these worst-case conditions, it most likely will perform well under normal conditions.

Most environmental testing is done in a test chamber. The conditions in the chamber can be cycled multiple times to accelerate product aging, and the PCA can be tested with or without power and signal inputs applied to it and monitored during the test. Specific standards for environmental testing procedures have been established by organizations such as the American Society for Testing and Materials (ASTM), the International Organization for Standardization (ISO), Underwriter's Laboratory (UL), the Society of Automotive Engineers (SAE), and others. Using these standards ensures that the same methods are employed consistently from test to test and that the results can be compared accurately.

It is critical that a designer knows the environmental conditions under which a PCA will operate and, if required, tested. In most instances, field data are available that define the environment a product is likely to be exposed to during operation. These data must be analyzed and applied correctly during the design of the PCA and interpreted correctly to arrive at the proper test environment to which it will be exposed. Using improper parameters can result in a product that may be either overdesigned or experience premature failure during operation. Either condition can have a major negative impact on the success of a design and the health of the company producing it.

5.5 PCA Reliability

A product's *reliability* generally is defined as the prediction of its ability to function properly for some specified period of time under specified conditions. Therefore, before starting to design a PCA, its reliability-related requirements should be identified. These normally should include

- Expected length of failure-free service
- Operating environment and conditions
- Acceptable performance parameters
- Expected reliability (failure rate as a function of time, such as mean time before failure)
- Impact of degradation in performance or failure on the rest of the system
- Maintenance and service approach to be used

The materials and components included in a design, along with the processes required to manufacture, assemble, and test it, affect a PCA's ultimate reliability. How an assembly will be packaged, stored, and shipped also need to be considered.

5.5.1 Failure Mechanisms

PCA failures usually can be traced to one or a combination of three sources: thermal stresses, mechanical stresses, and electrochemical mechanisms. These conditions may occur during manufacturing, operation, and shipping and storage of a PCA.

Thermal stresses are produced by exposure to temperature extremes, thermal cycling, and thermal shock (due to rapid changes in temperature). Thermal stresses can cause printed circuit board failures (delamination, plated hole damage, conductor breaks), component failures (either internal or package), open solder joints, and conformal coating/solder mask deterioration.

Mechanical stresses usually result from external vibration or shock inputs to a PCA. They may cause conductors to crack, component leads to break, and solder joints to fail. Parts may be overstressed and fail or break loose from a board.

Failures due to electrochemical mechanisms generally are initiated by exposure to humidity, corrosive materials, or contaminants, often a combination of all three. They can cause gradual degradation of the electrical characteristics of a circuit board (insulation resistance, characteristic, impedance, dielectric constant), cause corrosion, and induce metal migration and conductive whisker growth.

Unfortunately, the complexity and density of today's PCAs magnify the effects of these failure mechanisms. It is therefore critical that the designer, when making key layout decisions, considers the potential impact of these decisions on the ultimate reliability of the end product.

5.5.2 Influence of Design on PCA Reliability

There are many features of a PCA design that directly affect its reliability. They can be grouped in three main categories: the circuit board design features, the components used, and how the components are placed on the board.

Circuit board features that have an impact on reliability include

- *Interconnections.* Maximizing trace widths and spacings, where possible, to reduce performance degradation or failure due to processing defects, minimize electromigration, and avoid circuit performance failure due to current leakage or voltage breakdown.

- *Component land patterns.* Use pad configurations and locations that maximize solder joint quality, are compatible with standard manufacturing and cleaning processes, and are suitable for testing and rework or repair.

- *Vias.* Minimize aspect ratios and maximize diameters to reduce processing- and thermally related failures and optimizing barrel plating thickness to prevent cracking and edge discontinuities and tenting to prevent entrapment of corrosive materials and unwanted solder fill.

- *Board construction.* Alternate trace direction on adjacent layers and balance the number of layers around a board's neutral axis to prevent mechanical stress buildup, due to warpage, that could damage components, plated holes, conductors, or solder joints.

Component selection was discussed earlier in this chapter, but some additional reliability-related factors should be considered. These include

- Value tolerances should ensure proper functionality over the entire range of environmental conditions.

- Ability to survive assembly, test, and rework/repair processes without degradation of performance.

- Availability of component reliability data that can be related to the steady-state and cyclic stresses the part may be exposed to during manufacturing, operation, and servicing.

- Specification by the vendor of maximum allowable internal junction temperature (peak/steady state), junction-to-case temperature rise data, and derating recommendations.

- Susceptibility to degradation or failure due to exposure to electrostatic discharge (ESD), radiation, altitude, vacuum, or stresses caused by sudden environmental changes.

- Sensitivity to specific materials, chemicals, gases, or atmospheric conditions.

- Ease of assembly, test, and rework/repair.

- Compatibility with the circuit board design configuration and materials used and with the other components on the board.

- Use of new, special, or custom part types with little or no application reliability history.

- Use of obsolescent part types.

- Use of mechanically adjustable or variable parts.

- Use of plugin parts.

Where and how parts are mounted on a board can have a major effect on a PCA's ability to function reliably. Considerations include

- *Circuit performance.* Group functionally related parts to enhance signal integrity and prevent interference and degradation by signals from other circuitry.

- *Thermal layout.* Locate thermally sensitive parts for optimal heat removal and protection from external temperature influences.

- *Parts placement.* Place and orient parts to optimize manufacturability, testability, and ease of rework/repair.

- *Placement methods.* Use methods for mounting parts that minimize the effects of mechanical stresses.

- *Clearance and spacing.* Provide adequate clearance and spacing to avoid entrapment of contaminants and allow the use of proper cleaning processes.
- *Accessibility.* Allow room for visual inspection.
- *Accuracy.* Correct positioning of components demands accuracy.

5.5.3 Reliability Testing and Analysis

The main purpose of reliability testing is to gather sufficient data to predict a product's life-cycle performance. This is usually done by overstressing an assembly and compressing its time of operation to accelerate potential failure mechanisms. The data obtained from this accelerated testing are then extrapolated to actual operating conditions and used to predict an assembly's reliability. Potential weaknesses can be identified in this manner and the design corrected before a product goes into production or is put into operation.

The failure history of a PCA usually can be plotted on a bathtub-shaped curve, as shown in Fig. 5.6. Most electronic assemblies experience a high failure rate early in their operational lifetime due to "infant mortality." This is generally caused by a marginal part, circuit board, or solder joint that was not caught by production test or inspection but failed after a very few stress cycles.

5.6 Maintainability and Repairability

Most electronics equipment is built to function for years and must be maintained during its operational life. An important part of a designer's responsibility is to know how a PCA will be serviced once it is put into operation and to accommodate that requirement in the layout. Some PCAs are installed in products that are not intended to be serviced, but if corrective or preventive maintenance is required, it is frequently done in the field by personnel having a broad range of skill levels and using a variety of procedures and equipment. Consideration therefore should be given to

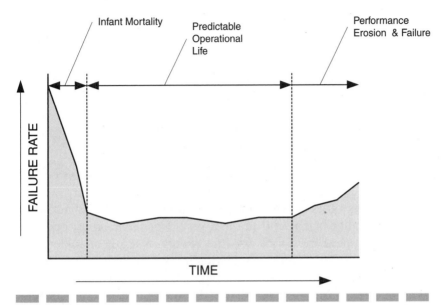

Figure 5.6
Reliability/failure history of a typical PCA.

providing in-place accessibility to test points and adjustments and allowing ease of removal and replacement of the assembly, if required, for more extensive test, adjustment, or repair.

5.6.1 Logistics

Maintenance can take many forms, ranging from local test, adjustment, or repair to returning the product to the manufacturer's facility. In most cases, simple servicing is done at a customer's site, whereas PCAs with complex problems are dealt with at a specialized repair facility or returned to the factory.

The skills of field service personnel may not be very high, and they may use obsolete or inadequate equipment to perform product maintenance or repair. It is incumbent on a designer to make the assembly as robust as possible to reduce the possibility of damage caused by servicing. Also, field service people usually have access to a limited inventory of replacement parts, so using as few unique parts as possible in a design will enhance its maintainability.

5.6.2 Rework and Repair

Three major types of rework or repair are usually performed on PCAs. They are

1. Correction of defective solder joints

2. Removal and replacement of components

3. Modification of the circuit board

These operations can be accomplished using hand tools or automated equipment.

Many types of hand tools are available for rework or repair, but the quality of the end result is very dependent on an operator's skill. Typical hand tools used for rework or repair of solder joints or soldered parts include soldering irons (usually temperature controlled), heated tweezers, thermodes (heated bars for desoldering multiple component leads), hot air nozzles, and IR emitters. Selection of the proper tool depends on the operation to be performed, the space available, and the heat sensitivity of parts and materials involved. Figure 5.7 shows examples of some of these tools.

Use of automated or semiautomated equipment reduces dependency on operator skill, but flexibility and versatility may be sacrificed. These types of tools do enable the user to exercise very precise control over the rework or repair process, resulting in predictable and reliable results. When dealing with high-density, closely spaced features or components, fine-pitch surface-mounted devices, or heat-sensitive parts, the ability to apply a preprogrammed heating/cooling profile and accurately positioning the tool may become essential to successful completion of the process. However, since this type of equipment is much more expensive than hand tools and requires a much higher skill level to set up and operate, a cost versus quality required tradeoff should be done to determine if such an investment is warranted. Figure 5.8 shows an example of a programmable rework/repair workstation.

CORRECTION OF DEFECTIVE SOLDER JOINTS The following are common PCA solder joint defects:

■ Nonwetting or dewetting-partial adhesion of molten solder to a surface

Solder wicking braid

Tweezer

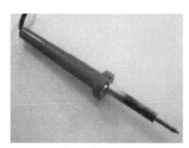

Soldering Iron

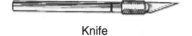

Knife

Needle nose plier and side cutter

Figure 5.7
Some typical hand tools used for rework and repair.

Figure 5.8
An example of a programmable PCA rework/repair
station. (*Used by permission of Pace USA.*)

- Insufficient solder
- Solder bridging
- Excess solder
- Solder icicles
- Voids, pits, pinholes in surface
- Cracked joints
- Granular or rough surface

Many of these defects can be corrected, but it should be noted that reheating a solder joint so that it can re-form itself uniformly is very difficult. In addition, the stresses caused by application of high, localized heat can damage adjacent electronic components and cause lifted pads, fracture of plated holes, and layer delamination on the circuit board.

When laying out a board, a designer can reduce the possibility of thermal damage caused by solder joint rework or repair by spacing components as far apart as possible and by providing thermal relief connections to large conductor areas, vias, and other pads. This tends to improve retention of heat at the joint being corrected and reduces the heating time required.

REMOVAL AND REPLACEMENT OF COMPONENTS The most critical part of changing soldered parts on a PCA is the removal of the component to be replaced. This is the part of the process that has the greatest potential for damaging the circuit board by the application of heat required to melt the solder joints that connect the component leads or terminals to the board. Also, if the assembly is conformally coated, the process of removing that material prior to unsoldering a part may cause mechanical damage to the board. Use of thermal relief connections, as described in the preceding paragraph, will help mitigate potential damage due to thermal stresses.

The key to successful removal of a component is the rapid and uniform application of heat to its soldered leads. Once the joints are molten, the part should be able to be removed quickly without damaging adjacent components or the circuit board. This may be especially difficult to accomplish for multipin, surface-mounted components, since all the solder joints may not melt uniformly, resulting in the application of excessive heat to some joints while waiting for others to become sufficiently molten. The designer should be aware of this possibility when defining and interconnecting pad patterns and try to keep the solder surface areas for multileaded devices as uniform as possible.

Before a replacement component is attached to a board, excess solder is usually removed from the pads, the remaining solder is leveled, and additional solder is provided if required. The new part must be aligned precisely on the existing pad pattern and held in place while the leads are simultaneously heated to reflow the solder. Proper alignment is critical, especially for very small surface-mounted chip devices and large devices with many fine-pitch leads, and optical aids may be required to accomplish this activity successfully.

CIRCUIT BOARD MODIFICATION Rework or repair of a circuit board usually is done at a service center or in a factory environment where the proper equipment and skilled personnel are available. Removal of solder material causing shorts, repair of cracked or open traces, and reattachment of partially lifted pads can be accomplished using properly controlled processes.

Circuit board modifications also can be performed to incorporate circuit changes. Traces can be opened and plated holes removed, if they are accessible and if there is minimal danger that the process will damage adjacent features. Jumpers can be added to create new circuit paths. The PCA designer should be aware of areas of circuitry that potentially could be modified in the future and, where possible, design in accommodations for these changes by providing extra pads for jumpers, routing traces in accessible areas, and additional space around the circuit features that may be revised.

Documentation

Understanding the importance of data and documentation, as well as how to maintain and control them, is one of the keys to producing a cost-effective end product. In today's electronics industry, printed circuit assemblies (PCAs) require that structured and controlled data be used throughout the design, documentation, revision, and procurement cycles to maintain product integrity and above all customer satisfaction. Industry standards have been in place for many years and are well defined for specific categories of documentation. Both military and commercial industries have established standards, general requirements, and guidelines for their preparation and structure. Recently, the IPC—Association Connecting Electronics Industries began to revise and control the majority of guidelines for fully describing the end product and its related support drawings. Table 6.1 identifies the most common types of drawings and electronic media needed to support product procurement.

A typical printed circuit board documentation package consists of a board detail drawing, a master pattern drawing or copies of the artwork masters, an assembly drawing, a parts list or bill of material, and the schematic drawing. This chapter outlines the considerations, methods, techniques, and tradeoffs involved in creating the documentation and data needed to produce a PCA. Primary emphasis will be placed on the potential effects that the selection of documentation methods and alternatives can have on the cost and performance of the completed PCA.

The documentation methodology described here is based on the assumption that the schematic, parts list, and circuit board layout have been created using computer-aided tools or have been captured in computer-readable format. Although a manually prepared data package would contain the same basic manufacturing, assembly, and test information, its use should be limited to building small quantities of noncomplex PCAs, such as engineering prototypes. As stated in Chap. 3, the basic source of information for drawings and artwork tooling is the manual layout, and most of it can be prepared by making reproducible copies of the layout and the taped artwork produced from the layout.

TABLE 6.1

Commonly Used Drawings and Data Formats Supporting Board Fabrication and PCA Assembly and Test

| Drawing | Common Data Format | Supporting | |
		Fabrication	Assembly/Test
Master pattern	Excellon	NC drill NC route board outline	—
	Gerber plot	Circuit layers Solder mask Marking	—
Assembly		—	Solder paste
Master pattern	ASCII	Aperture list Drill bit data Readme.txt file	—
Assembly		—	Solder paste Pick and place data Parts list
Board detail	HPGL plot	Manual drill and machining Manual route board outline	—
Assembly		—	Manual part place- ment Mechanical assembly
Schematic	IPC-D-356 or GENCAM	Bare board test	In-circuit and func- tional test

6.1 Circuit Board Artwork (Conductors, Solder Mask, Stencils, Marking)

In reviewing the requirements associated with creating and generating artwork for fabricating a circuit board, the designer should first identify the categories of data needed, which together will make up a typical artwork set (see Fig. 6.1). The content,

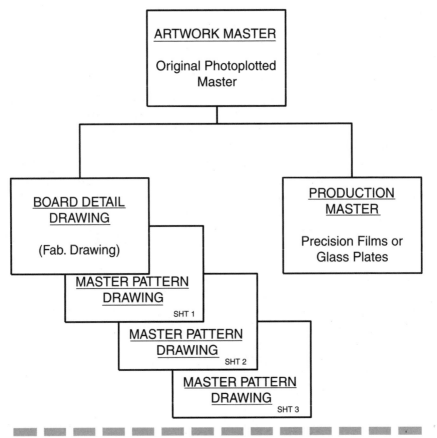

Figure 6.1
Example of a typical artwork set.

format, and structure of these data should be established, along with the methods to be used to identify and transmit them to a board fabricator. Compliance with industry standard format and film-generation process requirements and constraints will help ensure that the artwork (data or film) can be used by a supplier to produce good-quality boards.

6.1.1 Content, Format, and Structure

Artwork content must provide all the information and documentation needed for cost-effective bare board fabrication.

Table 6.2 lists the most common content of an artwork package with a cross-reference to current industry standards. When establishing artwork documentation content, format, and structure, the designer should tailor the data package to the specific needs of the fabricator that will be using the documentation to produce the circuit board.

A Readme.txt file, prepared in ASCII format, should be transmitted along with the data when transferring them electronically. This file will tell the board fabricator what data are included in the package and how they are structured and provide additional information such as company and product identification,

TABLE 6.2

Content of a Typical Artwork/Data Package for Board Fabrication

Artwork/Data	Format	Comment
Circuit layer 1	Gerber	Component side
Circuit layer 2		—
Circuit layer 3		—
Circuit layer 4		—
Circuit layer 5		—
Circuit layer 6		Solder side
Drill graphics		—
Board fabrication		Board detail drawing
Component-side solder mask		—
Solder-side solder mask		—
Component-side marking	↓	For screen fabrication
Solder-side marking	Gerber	For screen fabrication
Drill file	Excellon	NC drill data
Aperture list	ASCII	—
Net list	ASCII	—
Readme file	ASCII	Description of content of artwork/data package

support points of contact, purchase quantity, and quality certification requirements. Figure 6.2 shows a typical Readme.txt file that provides all the information needed for fabrication of a bare board.

IPC-D-350 defines the industry format standard (also known as *Gerber data*) that should be used when generating the master

---

readme.txt

Date: November 1, 1999
Time of transmission by modem: 0945 EST
To: Acme Board Fabricators, Inc, Attn J. Smith

Readme file for Widget Corporation Job No: 199-000079-001, Rev C
Purchase Order No. xab457321

Quantity required: 40
Delivery required: November 15, 1999

The file sent as 079001c.zip
Contents of file:
079001c.1	Gerber Artwork Layer 1 (component side)
079001c.2	Gerber Artwork Layer 2
079001c.3	Gerber Artwork Layer 3
079001c.4	Gerber Artwork Layer 4
079001c.5	Gerber Artwork Drill Graphics
079001c.6	Gerber Artwork Solder Side Solder Mask
079001c.7	Gerber Artwork Component Side Marking
079001c.8	Gerber Artwork Board Detail Drawing
079001c.drl	Excellon Drill File
079001c.apr	ASCII Aperture File
079001c.net	ASCII Net List
readme.txt	ASCII Text File (used to create this printout)

Please FAX confirmation of receipt of all listed files to A. Fisher @ 205-666-0789

Page 1

Figure 6.2
Example of a readme text file printout used to transmit data to a board fabricator.

pattern artwork plot files. The required artwork (layers), solder mask, screen, and stencil files should all use this format. The data identify final, to-scale image patterns, size and physical locations of all features, alignment targets, and markings. The file must be compatible with the software program in the plotter that will be producing the phototool films.

An aperture list is also supplied along with the artwork data. The list is the linkage between the plot commands in the artwork files and the photoplotter, telling it which patterns to draw on the film that will become the 1:1 master artwork. Most common aperture lists include an aperture number, shape, size, location, rotation, whether it is a draw or a flash, and whether it is to produce a positive or negative image. If a standard aperture library is established and coordinated between the designing and plotting functions, then this information can be provided as a position on an aperture wheel that contains predetermined aperture descriptions, thus eliminating the need for creating individual apertures for each new design. Artwork files, which identify the individual layer patterns to be used in the fabrication of the bare board, are normally structured and identified as being viewed from the component side through the board, as shown in Fig. 6.3.

An effective artwork file should contain the following items:

- Drawings and notes for processing controls
- Data format identification
- Board part number and revision
- Layer identification
- Quality conformance test coupons (as required)
- Stencil(s) (marking and paste) and solder mask(s)
- Imaging used (positive/negative)
- Aperture listing

The majority of today's circuit boards are designed on computer-aided design (CAD) systems. These systems incorporate the ability to extract data needed for fabrication of the designed board. On completion and verification of the design, drawing and artwork plot files can be created. The drawing plot files are used for board detail documentation and information. Artwork

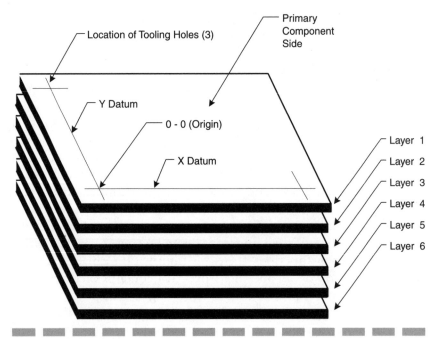

Figure 6.3
Artwork file structure.

files are produced for each of the individual board layers, screens, masks, and drilling references. In addition, a process support file is made, which contains the aperture information required by the photoplotter that produces the master patterns. The photoplotter creates precision 1:1 plotted artwork masters for

- *Production artwork masters.* Used for fabrication of the bare board, they are usually produced in compliance with the guidelines in IPC-D-310 on 0.007-in-thick, dimensionally stable polyester-type, photosensitive film or, in some cases where very high precision is needed, photographic glass plates.

- *Master pattern drawing.* Usually part of the board detail drawing, used in board fabrication to identify the required notes, board structure, dimensional profile, special machining requirements, and drilling references. Figure 6.1 shows the relationship between the master pattern drawing and the precision artwork masters.

In general, structuring of data consistent with industry standards broadens a company's access to external subcontractors that can cost-effectively fabricate complex circuit boards. If a company's capabilities allow it to make its own boards, the importance of structuring required data is less important. The majority of companies, however, rely on the board fabrication industry for support because the complex electronic products being produced today require the use of very costly equipment and controls.

When structuring data, the following should be considered:

- Media usage
  - Magnetic tape
  - Floppy disk
  - Cassette
  - CDs

- Fabrication needs
  - Drawings/notes/dimensional information
  - Gerber data/aperture list
  - Route/drill/bare board testing data
  - Stencils/screens

- Testing needs
  - Notes/test requirements
  - Net list

6.1.2 Methods of Data Transmission

Over the past 10 years, the time required for transmission of data to a board fabricator's facility has been reduced to just minutes, through the use of telecommunication techniques. However, schedule delays still result from such problems as lack of completeness of the information provided, improper data formatting, and incompatibility of interfaces between the sender and the receiver. For these reasons, the board shop's data content, structure, and format requirements must be identified and complied with to minimize impacts on cost and schedule.

After data files have been generated, they should be reduced in size by compressing them, which will speed up transmission and minimize the possibility of data dropout during transmittal. Methods currently being used for electronic transmission of circuit board fabrication data to suppliers include

- Direct modem-to-modem connection
- E-mail
- Posting to a Web page

If errors in the transmitted information are identified at the board supplier's facility, the same method should be used for retransmitting corrected data. Schedule impacts sometimes can be avoided by having the supplier correct the data. However, a copy of each affected file should then be sent back to the designer for confirmation that the change was made correctly and for insertion into the design database so that both sets of data are consistent with each other.

6.1.3 Data and Documentation Format Standards

Over the past several years, industry-wide standards have been established that define the format and structure to be used for electronically transmitted circuit board fabrication data. The IPC has been and continues to be the guiding force behind the development, publication, and dissemination of these standards, and their broad-based acceptance and use by both designers and manufacturers is helping to ensure the integrity and compatibility of electronically transmitted data. Among these standards is the IPC-D-35X series, which provides the guidelines for format compatibility while minimizing data redundancy. The series includes

- IPC-D-350, Printed Board Description in Digital Form
- IPC-D-351, Printed Drawings in Digital Form
- IPC-D-352, Electronic Design Data Description for Printed Boards in Digital Form
- IPC-D-354, Library Format Description for Printed Boards in Digital Form

- IPC-D-355, Printed Board Automated Assembly Description in Digital Form
- IPC-D-356, Bare Substrate Electrical Test Information in Digital Form

In addition to these electronic format standards, there are several others that define, identify, and control end product documentation. Typical information and guidelines that they provide include

- Documentation requirements
- Design, acceptability, and control
- Drawing formatting and structuring
- Marking and identification
- Dimensioning and tolerancing
- Fabrication, assembly, and testing
- Material and plating

6.1.4 Test Coupons

The conformance of a finished board to its design requirements is evaluated by examination of a test coupon that is associated with the board and exposed to the same fabrication processes. The configuration of this quality conformance test coupon is determined by the design configuration of the board and is usually included as part of its artwork. A board fabricator may include additional coupons on a panel as a way of evaluating and controlling the manufacturing process.

Guidelines for the content, construction, and location of test coupons are included in IPC-2221/2222, which define the configuration of various coupon sections that can be used for destructive and nondestructive tests without affecting the actual board. A quality conformance test coupon should provide the capability to evaluate every critical board characteristic (see Table 6.3).

A coupon must be identified with its associated board in order to ensure control of the test data. It should be labeled with the board part number and revision, lot date code, and manufacturer's identification.

TABLE 6.3

Typical Construction of Quality Conformance Coupon

Section	Description	Quantity per panel Class (see Sec. 1.1.2)		
		1	2	3
A	Evaluate solderability of plated through holes	N/A	2	2
B	Plating thickness	2˙	2˙	2˙
B	Thermal stress	2˙	2˙	2˙
B	Type 1 bond strength	2˙	2˙	2˙
C	Plating adhesion	N/A	1	1
C	Surface solderability	N/A	1	1
D	Interconnect resistance	N/A	1	1
D	Shorts	N/A	1	1
D	Circuit continuity	N/A	1	1
E	Moisture and insulation resistance	1	2	2
F	Registration	N/A	1	1
G	Solder resist adhesion	1	1	1

˙Opposite corners of panel.

6.1.5 Tools, Techniques, and Processes for Generating Artwork

Data used for generating precision artwork is usually produced with computer-aided tools, so the following description will be limited to that methodology (the general process for manual artwork preparation is described earlier in this book). The tools used to generate and flow down the data and ultimately create the artwork are:

- A computer-aided engineering (CAE) system to design the circuitry
- A CAD system to create the physical design

- A computer-aided manufacturing (CAM) tool to postprocess the design data

- A precision photoplotter to generate the 1:1 artwork required for board fabrication

Design engineers use CAE tools to symbolically define the components that are to be used and describe how they are to be interconnected. CAE systems can provide the following types of data and information:

- Schematic drawing
- Parts list
- Net list
- Test point data
- Results of circuit simulation and analysis
- Design rules and restrictions

The schematic, along with the parts list and the net list, contains intelligence regarding the properties of the parts and how they are connected. This aids in part placement and interconnection routing during PCA layout.

A CAD system uses the CAE outputs to produce a physical layout. CAD tools can provide the following types of data and information:

- Part placement
- Interconnection routing
- Bill of material
- Verification of interconnection spacing and continuity
- Confirmation that design rules and restrictions were followed
- Manufacturing drawings
- Manufacturing files (artwork, drill/machining data, assembly and testing data)

The manufacturing data files provide the information that a CAM tool needs to create inputs to a photoplotter and to provide additional support files required for fabrication of a circuit board.

CAM tools are basically tailored CAD systems with capabilities focused on fabrication, assembly, and testing processes.

Relative to artwork generation, a CAM system can

- Produce artwork tool files that include test patterns, panel borders, optical alignment fiducials, plating thieves, adhesive dams, etc.
- Create panelized circuit pattern files
- Add compensation to the artwork for various chemical and plating processes
- Generate photoplot data files
- Generate drill and route template files

On completion of the CAM process, the final data are transferred to a precision photoplotter. Information provided to a plotter usually consists of

- Plot commands to create artwork that will be used for producing
  - Individual conductor layers
  - Solder masks
  - Solder paste
  - Marking
  - Adhesive dots for attaching chip components
- An aperture list that provides information for
  - Feature shape size and rotation
  - Conductor width and spacing
  - Text and symbols

6.2 Drilling and Machining Data

The purpose of drilling a printed circuit board is basically twofold. One is to produce an opening through the board that will permit a subsequent process to form an electrical connection between conductor layers, and the other is to allow through-board mounting of leaded components. Different types of drilled holes, produced at various points in the board fabrication process, are required for unplated or plated through-holes for leads and through-the-board, buried, or blind vias. Figure 6.4 shows some of the different types of holes that can be drilled in a circuit board.

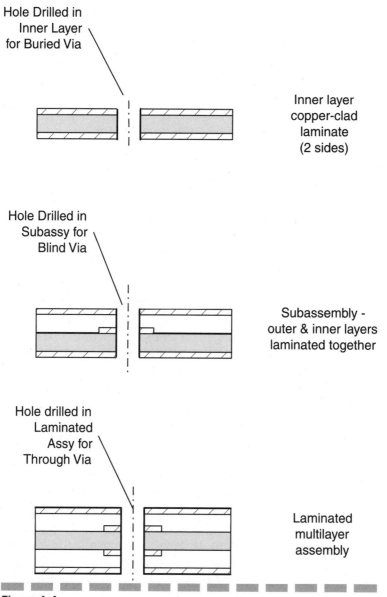

Hole Drilled in
Inner Layer
for Buried Via

Inner layer
copper-clad
laminate
(2 sides)

Hole Drilled in
Subassy for
Blind Via

Subassembly -
outer & inner layers
laminated together

Hole drilled in
Laminated
Assy for
Through Via

Laminated
multilayer
assembly

Figure 6.4
Different types of holes that can be drilled in a circuit board for vias.

Machining data provide the information needed to define mechanical features of a circuit board. Most fabricators use a routing process to establish a board's outline, employing either manual or numerically controlled (NC) equipment. If features such as pockets or slots are required, these also can be routed or, as an alternative, milled out.

6.2.1 Content, Format, and Structure

Three different types of media—board detail drawings, precision 1:1 artwork masters, and electronic data files—commonly define drilling requirements. All these are extracted from the CAD design system used to create the printed circuit board (PCB) layout.

Data contained in the board detail drawing include informational notes, pictorials, detailed views, and a drilling or hole chart. These document the requirements necessary to fabricate and drill a bare board. Notes in this document specify finished hole requirements. Pictorials and supporting detailed views identify board outline, dimensional tolerances, processing stackup(s), datums, and if a grid system is used for placement, the locations of off-grid holes. A drilling or hole chart provides information needed to identify hole sizes, tolerances, quantities, and plating requirements. All the information can be provided and controlled within one master fabrication document. When specifying drilling requirements for multilayer boards with blind or buried vias, the holes to be made in each circuit layer must be described separately. This also applies when providing drill data files for artwork plotting or to drive NC drilling equipment.

Precision 1:1 artwork contains the same information as the pictorial, although when processed the completed 1:1 artwork only displays graphic symbols, each representing the location and size of a hole to be drilled. The data needed to generate this separate artwork are derived from the layout in the circuit board CAD system. The drill pattern artwork also can serve as an overlay tool to verify that the drilling operation has been done correctly. Artwork plot data are commonly generated in the same format as that used for the conductor layer plots.

A drill data file provides the actual information that controls the drilling equipment used during board fabrication. Although a wide variety of NC drill equipment is used throughout industry, a de facto data format (Excellon) exists that is used by most companies; however, before generating and transmitting a drill file to a board shop, the designer should determine what format is compatible with the supplier's equipment. Figure 6.5 displays a typical multilayer artwork package with blind and buried via requirements.

Board machining information is provided in a similar manner. Fabricators that employ manual pin routing will use the dimensional data provided in the detail drawing to fabricate a template to profile the board and to manually machine any

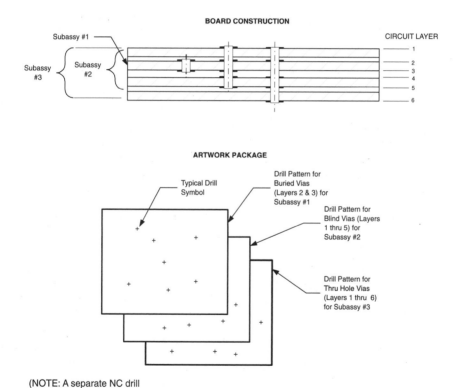

Figure 6.5

Artwork package for drilling blind, buried, and through-hole vias in a multilayer board.

other mechanical features. Board suppliers using NC routing equipment will require a separate file containing board profile data, provided in a format that is compatible with their equipment's controller.

If a board fabricator employs a CAM system to create phototool artwork and create files that drive its automatic equipment, then the designer should make sure that the drill and route data files are readable by that system. NC data can then be extracted, formatted, and input directly into the equipment controllers. Some of the more common capabilities of NC equipment used throughout the industry are identified in Table 6.4.

6.2.2 Methods of Data Transmission

Drill and route data files are transmitted by the same methods used for transmitting artwork data.

6.2.3 Industry Format Standards

The standards currently used throughout industry for drill information are

■ Artwork data are created as described in IPC-D-350.

TABLE 6.4

Typical NC Drilling Equipment Capabilities

Capabilities	1 Spindle	2–3 Spindles	4–5 Spindles	5–6 Spindles
Table speed	1000 in/min	2000 in/min	400 in/min	1000 in/min
Drilling accuracy	±0.0007 in	±0.0008 in	±0.001 in	±0.0007 in
Position accuracy	±0.0002 in	±0.0002 in	±0.0002 in	±0.0002 in
Position repeatability	±0.0001 in	±0.0001 in	±0.0001 in	±0.0001 in
Drill bit sizes	0.0039–0.250 in	0.0039–0.250 in	0.0039–0.250 in	0.0039–0.250 in
Tools per station	240	110	120	120
Maximum panel size	24 × 30 in	28 × 28 in	24 × 30 in	22 × 30 in

- Data that will provide commands directly to NC drilling equipment generally are supplied using the de facto industry standard, namely, Excellon format.

- General information is provided using text files.

The requirements for producing drawings and data are described in the following specifications and standards:

- Mil-Std-100, Engineering Drawing Practices (for U.S. government programs)

- IPC-D-325, Documentation Requirements for Boards, Assembly, and Support Drawings

- ANSI-Y14.1, Drawing Sheet Size and Format

- ANSI-Y14.5, Dimensioning and Tolerancing

6.2.4 Tooling Documentation

The board fabricator generally defines the tooling requirements for a panel, staying outside the board outline. Tooling features are added by manually preparing panel phototool artwork or by using a CAM system to generate a set of panel data files containing the following information:

- *Tooling holes,* which are used for board fabrication but also may be employed for alignment with assembly and test fixtures

- *Fiducials,* which are optical marks used for controlling and aligning stencils, screens, and masks, as well as for equipment with optical locating capabilities required for placement of fine-pitch parts

These tooling features are in addition to those provided within a board outline, which must be identified in the board detail drawing and master artwork. The general guidelines for locating tooling holes and fiducials are described in Figs. 4.5 and 4.7.

6.3 Bare Board Fabrication Data

The information needed for fabrication of a bare PCB should include the following general types of data and documentation.

They are defined and controlled by the board detail drawing (or as it is identified in IPC-D-325, the master drawing).

■ Dimensioned board details

■ Cross-sectional construction (layup)

■ Materials required and their associated specifications

■ Conductor definition and minimum spacing requirements

■ Fabrication allowances

■ Solder mask, screen, and stencil requirements

■ Marking requirements

■ Test coupon requirements

■ Performance requirements (mechanical and electrical)

■ Electrical test requirements

■ Master pattern drawing (this is usually part of the board detail drawing but can be a separate drawing)

■ CAE, CAD, and CAM reference data files

6.3.1 Content, Formats, and Structure

A board detail drawing usually includes these areas of information:

■ Notes

■ Board profile

■ Master pattern artwork

■ Drawing identification and control (title block, drawing number, revision block)

Notes should be placed on the first sheet of the drawing and should be clear, concise, written in the present tense, and capable of only one interpretation. The notes should define the following requirements for a typical PCB where applicable:

■ Finished board specification

■ Board material specifications (laminate, conductor, B-stage)

■ Type of board construction

- Solder mask material specification, thickness, and placement and feature constraints
- Marking content, location, and material specifications
- Drilled hole types, diameters, tolerances, allowable location error, etchback allowance
- Plating material specifications and thickness (in holes and on conductors)
- Finished conductor dimensions and allowable tolerances (including minimum annular ring)
- Minimum allowable spacing between conductors
- Allowable bow and twist
- Permissible processing allowances (artwork adjustment, modification of line/pad interfaces)
- Quality conformance test coupon location, marking, and test/inspection data to be provided
- Bare board test constraints and test data to be provided

If a circuit board requires special processing to incorporate cores, via tenting or filling, or special marking such as bag and tag identification, their requirements also should be included in the notes.

The board profile should be shown as it is viewed from the primary side ("main" component side). Any additional views needed for complete clarification of all mechanical details should be included. The information should specify board construction, overall thickness, dimensions, and tolerance requirements and define any special operations such as milling or additional plating. A typical board profile uses datums, which define the dimensional relationship between mechanical features (holes, edges, slots, etc.) and the datum intersection (origin). If a grid system is used to locate design features, the grid origin ("0-0 point") should coincide with the intersection of the datums. An example of a typical board profile is shown in Fig. 6.6.

A copy of each master pattern artwork required to produce a board is normally provided on the board detail drawing, including one for each of the individual conductor layers, sol-

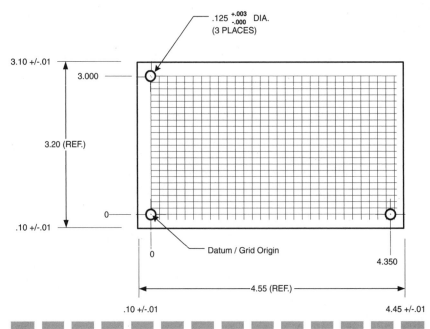

Figure 6.6
Example of board profile showing datum/grid origin.

der masks, markings, adhesive, and solder paste. These are all controlled by and referenced to the detail drawing. All artwork patterns are produced as viewed from the primary side of the assembly.

The support data most commonly provided are in an electronic format and used for drilling, NC routing and milling, plots, testing, and creation of the precision master artworks. These files are further defined in Table 6.5 for reference.

6.3.2 Material and Layer Construction

The board detail drawing specifies the materials and construction required for the bare board. The materials typically are identified within the notes and are cross-referenced to a detailed construction view. These notes define the requirements for

■ Laminate dielectric material

■ B-stage

TABLE 6.5

Support Data Most Commonly Provided in Electronic Format

File Type	Typical Information Provided
Pin locations	Part reference designators, device types and symbol names, pin symbol names, pin numbers, pad stack names, net names, XY locations
Component locations	Part reference designators, device types and symbol names, package symbol names, origin XY locations, orientations
Bill of material (BOM)	Device types, package symbol names, part reference designators, quantities
Test points	Net names, XY locations, surface locations
Unconnected pins	Part reference designators, pin numbers, pin symbol names
Drill data	Drill sizes, XY locations, quantities, drill routines (plunge, etc.)
Net report	Net names, part reference designators, pin numbers
Artwork data	Aperture sizes, D-code locations, rotations, flash/draw designations, layer allocations
Board data	Physical features (outline, tooling, machining, etc.), XY locations, construction
Summary report	Design rules, padstacks, trace/pad dimensions, tolerances

- ■ Copper foils
- ■ Cores
- ■ Masking materials
- ■ Marking materials

Construction views in the board detail drawing identify the sequence of materials that make up a board. This detail defines the material stackups, individual layer identification, and controlling dimensional requirements. Figure 6.7 identifies the contents and structure of a typical board stackup.

6.3.3 Drill Data

The drill data provided on a board detail drawing should include

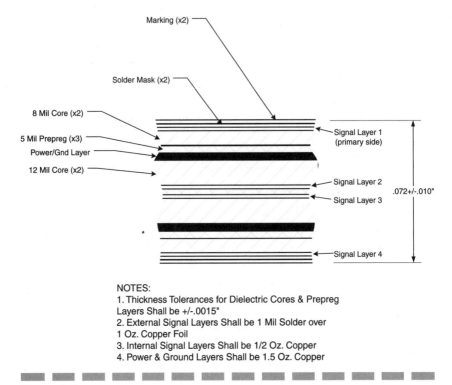

Marking (x2)

Solder Mask (x2)

8 Mil Core (x2)

5 Mil Prepreg (x3)

Power/Gnd Layer

12 Mil Core (x2)

Signal Layer 1
(primary side)

Signal Layer 2

Signal Layer 3

.072+/-.010"

Signal Layer 4

NOTES:
1. Thickness Tolerances for Dielectric Cores & Prepreg
Layers Shall be +/-.0015"
2. External Signal Layers Shall be 1 Mil Solder over
1 Oz. Copper Foil
3. Internal Signal Layers Shall be 1/2 Oz. Copper
4. Power & Ground Layers Shall be 1.5 Oz. Copper

Figure 6.7
Typical multilayer board material construction view.

- Hole size and tolerance requirements
- *XY* locations
- Plating requirements

This information is usually provided in the form of a hole schedule chart (see Fig. 6.8 for an example). The hole chart uses a unique symbol to define each different size and type of hole required. A plotted or drawn drill pattern shows the location of a symbol for each hole in the board (see Fig. 6.9). If blind or buried vias are required, involving sequential drilling operations in different layers or subassemblies and separate drill charts and associated drill pattern plots, identification of each individual operation should be provided. If a grid system is used to define the locations of parts, holes, and other features on a board, holes

	Description	Quantity
△	.028±.003 dia. Plated through	38
○	.035±.005 dia. Plated through	52
☐	.125 - .127 dia. Nonplated	3
✕	.056± .003 dia. Nonplated	16
◇	.042± .005 dia. Plated through	28

Figure 6.8
Typical hole schedule chart.

Each Symbol Shows
Location & Type of Hole
Required

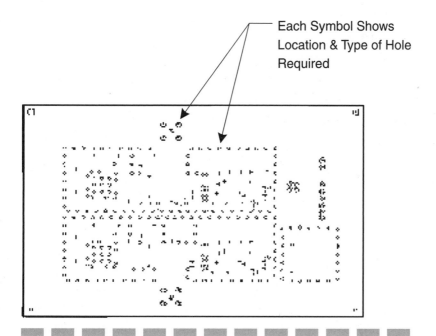

Figure 6.9
Circuit board drill pattern.

need not be individually dimensioned unless they are not centered on a grid intersection.

6.3.4 Board Dimensions

A board profile or outline view, with required dimensions, is an integral part of the detail drawing. It depicts feature sizes; their positions, locations, and forms, and associated tolerances. The electronics industry traditionally has used bilateral dimensioning and tolerancing to define physical board features; however, geometric dimensioning and true-positioning tolerancing are becoming more widely used because of the following advantages:

- They allow a greater percentage of tolerance area.

- They provide maximum producibility.

- The maximum/least material concept provides a wider tolerance allowance.

- They aid in automatic assembly.

- They ensure interchangeability of mating parts.

- They reduce controversy and guesswork by providing a uniform way of identifying and interpreting dimensional requirements.

Board dimensions define and control the overall board size, thickness, and special machining requirements. Tooling holes establish the board dimensions. They act as alignment guides for assembly and test fixtures. These holes are used to establish the design datum to which all dimensions are referenced. Figure 6.10 contains two typical board outline views, one using bilateral dimensioning and the other with geometric dimensioning.

6.3.5 Plating, Marking, and Masking

Usually a drawing will identify areas to be plated with references to notes for the actual plating requirement. Markings are also controlled this way unless the identification for marking is

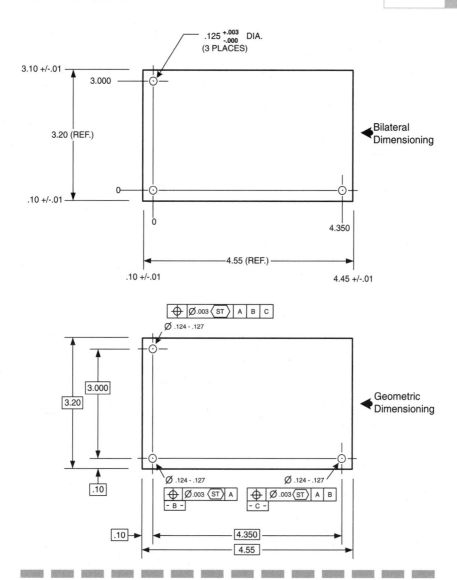

Figure 6.10
Examples of bilateral and geometric board dimensioning.

applied by screening, which requires creation of artwork for producing the screen. A 1:1 reproduction of this artwork is included in the detail drawing along with solder mask or solder paste artwork copies, if required. Figure 6.11 identifies a typical board marking and location requirements.

As a general rule, any board that has exposed external circuitry and is to be soldered using a mass process (wave, bath, or reflow) should incorporate a solder mask in its design to aid in the elimination of solder bridges. Solder mask materials may be applied either as a liquid or as a dry film. Depending on the application and processes, they both provide dielectric and mechanical shielding during and following soldering operations. The IPC-SM-840 specification provides guidelines, test methods, and conditions for the uses and types of solder masks.

Plating, marking, and masking requirements may be identified on a detail drawing by

- Notes
- Drawings
- Precision artwork masters
- Electronic data

Plating, marking, and masking notes should contain the following information:

- Plating
  - Types
  - Material specification
  - Location
  - Thickness
  - Surface condition
- Marking
  - Types
  - Material specification
  - Location
  - Color
  - Methods
  - Height and width of finished characters
- Masking
  - Material type

- Location
- Material specification
- Thickness

6.4 Assembly Data

An assembly drawing provides all the necessary detailed information required to manufacture a PCA. The data in an assembly drawing normally are derived from the board layout and its associated schematic. Individual part outlines and reference designators are shown pictorially on the drawing. Item find numbers are included in the component view, and they are cross-referenced to each unique part type (part number) in a bill of material or parts list. The bill should include the quantity required, reference designators, part number, and a brief but concise description for each unique part number.

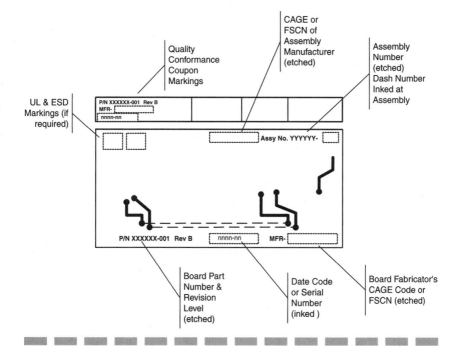

Figure 6.11
Typical markings on a bare circuit board.

Any mechanical hardware that attaches to or supports the PCB, such as stiffeners, brackets, ejectors, etc., is identified on the assembly drawing. The assembly drawing also may identify the locations of markings applied during the assembly operation. Generally, the type of marking and methods used will be defined in the notes section of the drawing.

6.4.1 Content, Format, and Structure

An assembly drawing must be consistent with all other documents and data that define the fabrication, assembly, and test requirements of a PCA. The linkage with drawings and electronic data files must be carefully cross-referenced and its accuracy verified to eliminate possible errors. The flowchart in Fig. 6.12 shows the main connections between an assembly drawing and the other PCA design data.

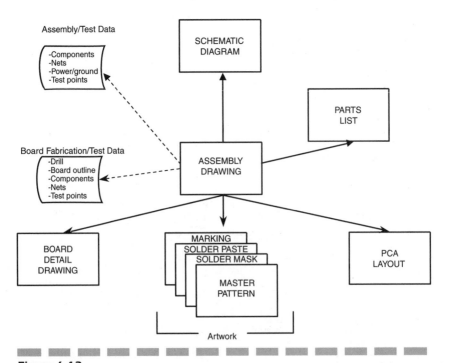

Figure 6.12
The main connections between an assembly drawing and other PCA data.

The primary elements included in an assembly drawing are

- The circuit board definition
- Physical location and mounting of electronic components
- Marking requirements
- Soldering requirements
- Masking requirements
- Conformal coating requirements
- Installation of mechanical hardware
- Notes
- Bill of material
- Electrical test requirements
- Quality requirements

Many general requirements can be defined in the notes section of an assembly drawing, including

- Workmanship
- Special handling
- Dimensional references
- Component mounting and orientation
- Material specifications
- References to other documentation

Since diverse organizations and resources are involved in manufacturing a PCA, it may be appropriate to have representatives from design, board fabrication, assembly, test, quality, configuration control, and standards functions agree on the content and structure of the assembly notes. This is especially appropriate if a family or group of PCAs is being designed for a project and a standard set of notes is needed. Some of the more common issues to be considered when establishing assembly notes are identified in Table 6.6.

6.4.2 Component Mounting

Individual electronic components are assigned reference designations (such as U1, C10, CR16, R58, etc.) that are either specified

TABLE 6.6

Typical Assembly Note Content

Type of Note	References
Drawing	Standards used, method of generation (CAD/manual), reference drawings
Workmanship	Compliance with designated specification or standard
Identification	Assembly number, board detail number, CAGE code, serial number, revision control, location/size/material for markings, marking standard or specification to be used
Special handling	Electrostatic discharge protection, if required, sensitivity to thermal or mechanical shock
Conformal coating	Material/application specification, thickness, noncoated surfaces
Dimensions	Tolerances, assembly envelope references, maximum/minimum allowances, dimensioning/tolerancing method
Component markings	Location/size/material, reference designators used/missing, spare part locations
Part mounting	Orientation/polarity/pin 1 indicators, general lead forming requirements, general spacing requirements
Assembly processes	Adhesive application, fastener/rivet/eyelet installation, assembly sequences, process specifications
Soldering	Material/application specification (including solder paste), solder placement, solder plug/masking requirements, cleaning procedures
Inspection/testing	Methods, process specifications, quality requirements

on the schematic or allocated during the layout activity. They should be assigned in numerical sequence for each type (such as integrated circuits, capacitors, resistors, diodes, transistors, etc.) and in ascending order. These reference designations must match those in the associated schematic and the parts listed in the assembly bill of material.

The location, orientation, and polarity of every component should be indicated on the assembly drawing. If multipin parts are used, one pin should be designated on the main view (usually pin 1). Lead forming requirements are to be specified, and component body spacing off the surface of the board must be defined. Detailed views of specially mounted parts (such as those requiring heat sinks, adhesive bonding, or special insulators or hardware) should be provided. Nonstandard lead bends or off-center component installations also should be detailed, unless the parts are procured with preformed leads. Areas of the assembly having height restrictions above or below the board should be identified.

Component lead extension after soldering is to be specified, and if clinched leads are required, these shall be indicated on the drawing. If point-to-point wiring (jumpers) is needed, its installation (routing and tie-down locations) is to be documented on the assembly drawing. Mounting of all other circuit elements, including bus bars, test points, terminals, eyelets, indicator lights, switches, connectors, and wire harnesses, should be described in detail. The assembly drawing should indicate by note or callout on the pictorial view any components requiring special handling during assembly, such as parts sensitive to electrostatic discharge (ESD).

6.4.3   Soldering Requirements

Solder materials and finished solder joint quality requirements are usually specified in the assembly drawing notes. If specific plated holes are to be plugged with solder or kept free of solder, these must be designated.

For assemblies needing application of solder paste, their location and specifications (material, thickness, etc.) should be defined. If parts are used that may be damaged by heat generated during mass soldering operations, these must be identified as requiring installation with individual mounting and interconnecting processes.

6.4.4   Mechanical Assembly

Most PCA designs contain some mechanical parts and hardware. Some of the more common types of hardware used for electronic assemblies are screws, nuts, washers, mounting clips, heat sinks, rivets, handles, ejectors, wedgelocks, stiffeners, and brackets.

The location and orientation of these items should be described on the assembly drawing. Mechanical forces associated with the mounting of hardware must be controlled carefully to ensure that there is no damage during attachment to a circuit board. Maximum allowable tightening torques and crimping forces must be provided as part of the assembly description.

6.4.5 Marking

An assembly should be marked in a way that provides traceability between it and the documentation that defines its configuration. Assembly part number marking can indicate variations of a basic design (by part dash number and/or revision letter) and also include a reference to a specific PCA's manufacturing and test history by serialization. Other markings, such as part reference designators, polarity or orientation indicators, ESD status, and others, are usually defined and provided on the bare board and not specified on the assembly documentation.

The materials and methods selected for marking must satisfy these purposes by being readable, durable, and compatible with the required manufacturing processes as well as the end use of the PCA. The assembly drawing should identify and control the materials, types, and locations for all markings applied at that level. It should be noted that if conductive ink is being used, it must be permanent and appropriately isolated by either spacing or a form of coating from the conductive circuitry. Table 6.7 identifies typical marking methods.

6.4.6 Protective (Conformal) Coating

There are two basic types of protective coatings applied to PCAs: solder mask and conformal coating. A solder mask is usually applied when fabricating a bare board, and the documentation defining its requirements is discussed in Sec. 6.3.5.

Conformal coating is a permanent insulating material that conforms to the surface of the object being coated. Conformal coatings should be homogeneous, transparent, and unpigmented. Such coatings usually are applied after a circuit board has been

TABLE 6.7

Typical Marking Methods

Marking	Commonly Used Marking Method/Process
Assembly identification:	
Root identifier (assay no.)	Etched, ink screened, or stamped on bare board
CAGE code	Etched, ink screened, or stamped on bare board
Assembly dash number	Ink screened or stamped at assembly
Assembly serial number	Ink screened or stamped at assembly
Component markings (reference designators, part location indicators, orientation/polarity/ pin 1 markings)	Etched, ink screened, or stamped on bare board
Electrostatic discharge (ESD)	Etched, ink screened, or stamped on bare board or decal/label applied at assembly

assembled and provide an excellent barrier against damage to the circuit due to moisture, chemicals, and abrasive materials. The basic types of conformal coatings used are ER (epoxy), UR (urethane), AR (acrylic), SR (silicone), and XY (*para*-xylene). These coatings may range in thickness from 0.5 to 3.5 mils depending on the type used.

The information that controls the material, thickness, and location of conformal coatings is provided by bill of material entry, notes, and callouts on appropriate assembly drawing views.

6.4.7 Parts and Materials Data

The bill of material (also called a *parts list*) is a tabulation of all parts and materials required to manufacture a PCA, including an entry for each unique part or material number, a succinct description, the quantity of each required, its item or find number, and for an electronic part, the reference designators used. Materials used during the manufacturing process should not be included. In addition, reference information may be listed, such as the schematic drawing number, test specifications, and applicable industry standards.

A bill of material may be an integral part of the assembly drawing or may be created as a separate document. If the bill is issued as a separate document, it should have the same root number as the assembly drawing (possibly with a "PL" prefix) and should be maintained at the same revision level. The general guideline for structuring this information is to group like items by type and place them in ascending alphanumeric part number order. Figure 6.13 shows examples of the structure of typical bills of material.

Each part used on a PCA should be identified somewhere in the field of the assembly drawing by an item or find number that corresponds with its entry in the bill of material, and each electronic component should be identified by its reference designator. Bulk materials, such as solder, marking ink, and conformal coating, may be identified in a note or on the field of the drawing. Reference designators indicated on the drawing must be verified carefully to ensure that they are consistent among the following documents and data:

- Bill of material
- Schematic and data extracted from it
- Circuit board layout and associated data
- Marking artwork
- Assembly drawing
- Assembly and test data

6.5 Testing Data

Circuit or electrical testing ensures that an assembly and its components conform to the specified circuit performance requirements. These requirements are usually identified in a system or subsystem specification and are flowed down and implemented by a PCA circuit design, which is documented on a schematic diagram. Information needed to perform the testing is mainly derived from the schematic and the PCA layout and augmented by a test specification that defines the test requirements, procedures, conditions, accept/reject criteria, and equipment constraints.

Company					Assembly P/N:		Revision Nbr:	Date:
Address					Title:		Change Tracking Nbr:	
Cage Code Nbr:					Contract Nbr:		Approval:	
Find Nbr	Qty	Cage Code	Part Number	Dwg Size	Nomenclature or Description		Company Part Number	

-002	-001	Find Ndr	Part Number	Cage Code	Nomenclature or Description	Material or Note	Company Part Number
Qty Required							

Figure 6.13

Two examples of bill of material structures.

6.5.1 Content, Format, and Structure

CAE and CAD tools of today provide an extensive source of information and data that can be used for circuit testing. "Canned" or preformatted system reports and data files can be extracted, and macros, scripts, and programs can be created that provide additional customized outputs that can be used for a variety of related activities.

Electrical testing can be done in four basic phases:

1. *Bare board test.* Verifies the circuit interconnections.

2. *In-circuit test.* Verifies the correctness of the circuitry and electronic components on the assembled board.

3. *Functional test.* Verifies that the PCA executes its intended functions properly.

4. *System integration test.* Verifies that the PCA operates properly when interfaced with the rest of the system.

The information commonly provided for testing is included in drawings, test specifications/procedures, or electronic files extracted from CAE and CAD design tools. Drawings and specifications generally are used to define test requirements, whereas electronic data files provide the information needed for producing test fixtures and programming the test equipment. Table 6.8 describes the contents and common uses of the information contained in drawings, test specifications, and electronic data files.

6.5.2 Test Point Locations

For bare board testing, which is done mainly to check for conductor path opens and shorts, the physical location (X, Y distance from the design origin) of test points is extracted from the layout. This information can be generated manually and provided in a test specification or produced by a CAD system as a data file to be used to program an automated tester. Locating test points for in-circuit testing is a more complex process, since both the interconnections and electronic components must be verified.

The optimal location for circuit test points is on the solder side of a PCB. This allows use of a single-sided test fixture and

Documentation

TABLE 6.8

Test Support Information Contained in Drawings, Test Specifications, and Electronic Data

Source	Contents	Information Use
Test specifications	Identification of standards and guidelines defining PCA performance, test requirements and test procedures	Information for bare board, in-circuit, and assembly-level testing Documentation/data references and linkages
Drawings	Master artwork	Physical location/orientation of circuit components Physical location/orientation of test points and connectors Physical information for procurement/build of test fixtures
	Board detail Assembly	Documentation/data references and linkages
	Schematic	Visual identification of signal flow, signal/component names, circuit test points and circuit requirements Documentation/data references and linkages
Electronic data	Physical reference designations of component and test nodes XY component pin and test point locations	Component and test locations for test equipment program generation Wiring of test fixtures
	Component and board circuit characteristics	Component, signal, and circuit criteria defining test program requirements

simplifies test equipment programming. Although the average complexity and density of today's PCAs make this almost impossible to achieve, maximizing the number of test points on the solder side will have a beneficial impact on the cost of testing.

Most CAD systems in use today provide a designer with the ability to optimize the locations of test points on a PCA. After components have been placed on a layout, an internal routine can be executed that identifies the connections (nets) of the circuit that can be tested automatically from the solder side of the assembly. This routine also can provide a means to determine which nets can be brought through from the component side by introducing a via and/or test pad into the circuitry and which nets require the addition of a test pad on the component side. Circuit elements can then be added as test point components that are interconnected during routing.

In order for this activity to occur, the land area or pad for a test point must be included in both the circuit element and physical libraries. Both libraries should contain two types of test points, which are independently identified as separate component parts. One part represents a solder-side test point and the other a component-side test point. Once a test point has been selected and placed as a part, the proper electrical symbol, based on its location in the circuit, must be added to the schematic.

From this information in the layout and schematic data files, an X, Y location and circuit function of each test point can be extracted and provided in a test data file to be used to build a test fixture and program a tester. Figure 6.14 shows an example of how circuit test points can be added and documented for an assembly.

6.5.3 Component Identification and Circuit Information

The identification (reference designator and pin number), X, Y location on the circuit board, and net association of each electronic component's terminal is required for in-circuit testing. This information is originally defined in the circuit and component libraries and extracted from the schematic and layout files. This information also may be needed for programming a functional tester.

6.5.4 Fixture Requirements

The documentation required for test fixtures generally consists of mechanical drawings and electronic data used to locate and identify the test probes that will make contact with the surface of a board during circuit testing. Three types of test fixtures are commonly used in testing:

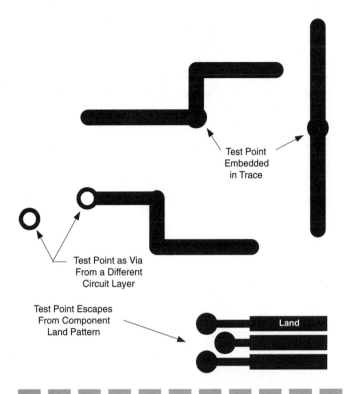

Figure 6.14
Typical circuit test point.

- *Bed-of-nails.* Bare board test
- *Single-sided.* In-circuit test
- *Double-sided (clamshell).* In-circuit test

The bed-of-nails type of fixture requires *XY* probe point location, part, and net identification and should be configured as described in IPC-ET-652. Test data are obtained from the PCA layout. Single- and double-sided test fixtures, used for in-circuit testing, are also defined by the layout, along with data from the schematic. Clamshell fixtures, also used for testing assemblies, provide the capability for testing both sides of an assembly simultaneously. Although very costly, this type of fixture is commonly used for testing assemblies with components on both sides of a board.

Another type of tester being used primarily for bare board verification incorporates moving probes to make contact with the surface of a board. It requires very minimal mechanical fixturing, but the testing cycle time is much longer, since each net must be probed sequentially. Figure 6.15 shows the construction of a typical single-sided bed-of-nails fixture used in the industry today for either bare board or in-circuit testing

6.6 Part and Documentation Numbering

The documentation package for a PCA is intended to define all its requirements and control its as-built configuration. To effectively accomplish this purpose, all elements, including drawings and data, should be identified and interrelated by using a numbering system that provides the proper linkages between each other and the hardware itself. In addition, the numbering approach must be able to accommodate and control changes to

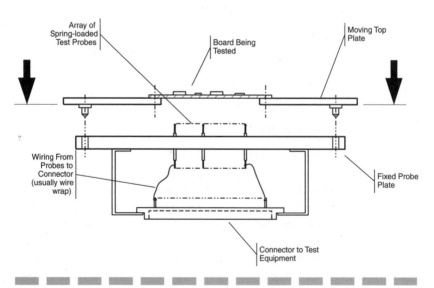

Figure 6.15
Construction of typical bed-of-nails test fixture.

the documentation and relate those changes back to the physical and electrical configuration of the PCA.

Defining the relationships between the various types of documentation needed to control a product provides the basis for creating a drawing and data numbering system that reflects these relationships and allows the data to be managed and used more easily. Table 6.9 shows an example of how such a numbering methodology could be implemented. In establishing an identification and numbering structure, the following data elements should be included:

- Drawings
  - Assembly
  - Components
  - Board detail
  - Specifications
  - Bill of material
- Data
  - Schematic
  - Layout and artwork
  - Fabrication
  - Tooling
  - Assembly
  - Testing
  - Archived material

These can be broken down into subcategories that may be needed to support fabrication, assembly, test, and integration of the end product.

6.6.1 Part Identification

There are several types of parts used in a PCA design that may require implementation of documentation controls. These are identified in three major categories: purchased, purchased and modified, and custom-produced for a specific design. Within these three categories may be both electrical and mechanical parts.

These parts are identified in the schematic and/or the physical layout and all associated documentation for a PCA. The

TABLE 6.9

An Example of a PCA Data/Documentation Numbering Structure
Numbering Structure

Drawing	Category Number	Base Number	Configuration Control Number	Defines
Schematic	100	000100	0001	Circuit requirements Design data base Simulation/test data Procurement data
Master pattern	101	000100	0001	Artwork data for board fabrication Artwork data for solder mask and marking
Board detail	102	000100	0001	Drill/profile/machining data Data for fabrication of test fixtures
Assembly	103	000100	0001	Part pick/place data Procurement data Mechanical assembly Performance requirements

Notes: The category number identifies the type of data/documentation. The base number identifies an individual PCA. The configuration control number shows the data/documentation change history (mainly revisions that affect form/fit/function of the PCA).

information associated with each part used in a design is derived from the circuit symbol library and the physical part library. To control the identification and technical content of the descriptions in these libraries, a consistently applied, rules-based part numbering system should be used within a company. This will aid in the identification of similar parts, prevent duplication of library entries, and move an organization toward standardization of parts use in new designs. Establishment of a uniform device naming/numbering convention or *key*, that associates each library entry with a part family or grouping, as well as identifying its part number designation, will enhance a designer's ability to find and select parts for a design. The key provides a linkage that is used through the various design tools and end product documentation as a common reference to the part. This approach provides consistency and verification of correct part utilization and designation within the following areas:

- *Part drawing.* Procurement data controls
- *Circuit symbol library.* Part entry controls (schematic and parts list)
- *Physical shape library.* Part entry controls (layout)
- *Bill of material.* Control of supplier part numbers
- *Analysis.* Control of analysis/simulation models (behavioral, timing, signal integrity, thermal)

If carefully designed during a company's planning for identification and control of parts, the key structured part identification system will enhance the effectiveness and efficiency of library management activities and will aid in standardizing part utilization in PCA designs. Figure 6.16 identifies the flow of part information throughout the design process. Table 6.10 shows typical linkages and controls associated with part data.

6.6.2   Documentation Identification

In order to establish a documentation numbering structure, the basic data elements that will be needed to produce and support the end product should first be identified. The following general types of data usually need to be identified and linked to each other:

- Analysis
- Design
- Part procurement
- Tooling
- Fabrication
- Assembly
- Testing
- Maintenance and repair

Table 6.11 shows individual activities within each of the preceding categories, along with their associated data outputs.

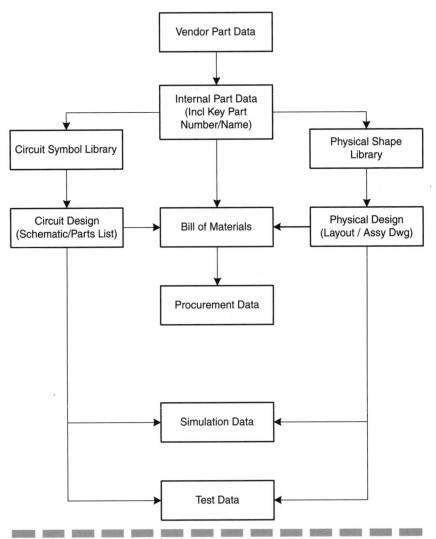

Figure 6.16
Flow of component part information during the design process.

DRAWING IDENTIFICATION All the drawings that are generated during the PCA design process must be identified. A company's internal engineering practices usually will define the structure, controls, methods, and requirements for drawing identification. All drawings, regardless of their individual function, should use common identification characteristics, including

TABLE 6.10

Linkage and Control of Part Data

	Main Linkages	
Part Data	**From**	**To**
Supplier/vendor part number	Part data	Bill of materials (BOM)/parts list (PL)
Standard circuit performance	Part data	Circuit symbol library Circuit simulation model Test programs
Mechanical description	Part data	Physical shape library Thermal simulation model
Internal company part number	Company part numbering convention	Circuit symbol library Physical shape library BOM/PL
Circuit symbol/ physical shape names	Company symbol/ shape naming convention	Circuit symbol library Physical shape library
Circuit symbol	Circuit symbol library	Schematic
Physical shape	Physical shape library	Layout
Part application in PCA design	Schematic Layout	Simulation models (performance) Test programs BOM/PL (quantity)
Part reference designator	Schematic Layout (back annotation)	Layout Schematic (back annotation) Assembly drawing BOM/PL

- Title
- Drawing number (root number)
- Revision control
- Linkage to other drawings

A drawing's title provides a name by which a part or item will be known. Some drawing titles require modifiers to differentiate between similar items used in the same assembly. Identifiers for drawings should not exceed 15 characters. Beyond this, the numbering system can become unwieldy. Within a data set, such as

TABLE 6.11

General PCA Data Requirements

Type of Data	Activities	Outputs
Analysis	Circuit simulation/modeling Breadboarding Thermal/vibration modeling	Test reports
Design	Circuit design PCA design	Schematic Layout
Part procurement	Identification of part numbers and quantities Definition of custom, semicustom, and modified parts requirements	Bill of materials Part specifications
Tooling	Provision of fabrication, assembly, and test tools and fixtures	Assembly and detail board drawings/files, artwork for solder paste/marking/solder mask
Fabrication	Board fabrication Mechanical part manufacturing	Detail board drawing/files, master artwork
Assembly	PCA assembly	Assembly and detail board drawings/files, parts list
Testing	Bare board testing In-circuit testing Functional testing	Test data (circuit, physical design files)
Maintenance and repair	Adjustment Part replacement	Drawing set, field manual, standard repair procedures and tools

that which comprises the detail drawing, there may be a need to add dash numbers to the root identifier to provide unique configuration descriptions for a family of parts that have minor variations. Using dash numbers can minimize the amount of documentation required to define such a family of parts or assemblies.

A revision designator on a drawing is used as an indication that the drawing has been revised after it has been reviewed, approved, and put under document control. Revision letters are used most commonly for controlling and identifying drawing changes. A typical method for using revision letters is identified in IPC-D-325, Documentation Requirements for Printed Boards, Assemblies, and Support Drawings. Linkage between drawings and data also should be identified. These links generally are defined in a drawing's notes and in an application block for an item's next-higher assembly.

MARKING OF PARTS AND ASSEMBLIES Markings should be used to provide reference designators, part and serial numbers, and revision levels. Fixed information, such as part numbers, should be incorporated in the master artwork. Test coupons should be marked with the same number as the circuit board for traceability. Variable-format information, such as serial numbers, fabricator information, date codes, etc., should be placed in an appropriate area and identified by using permanent nonconductive high-contrasting inks, labels, laser scribes, or other means that provide sufficient durability to survive assembly and cleaning. The detail drawing should indicate locations for these traceability markings.

Assembly marking should include CAGE (commercial and government entity) code or a federal stock certificate number (FSCN), assembly number, dash number, revision, and serial number. The CAGE code or FSCN and assembly numbers can be identified in copper or permanently ink screened. The dash number, serial number, and revision should not be identified in copper or permanently ink screened. These items should be marked at the time of component assembly and usually are hand stamped.

DESIGN DATA FILES Three interrelated databases are created and maintained during the PCA design process. They are

■ *Circuit database.* Schematic data
■ *Mechanical database.* Assembly data
■ *Design database.* Circuit board layout data

The circuit database contains the schematic drawing and provides information for the following files:

- Circuit netlist
- Bill of materials
- Testing data
- Circuit analysis

The database is identified using the same number and revision level as the schematic drawing.

The mechanical database contains the assembly drawing and provides information for the following files:

- Piece part drawings
- Assembly processes
- Bill of materials
- Subassembly drawings

The data structure for the mechanical database is the same as the circuit database, except it is identified with the assembly drawing number and revision level number.

The design database contains the layout drawing and provides additional information, which supports the following files:

- Bare board fabrication data
- Assembly data
- Testing data
- Test tooling data

The data structure, although similar to the circuit and mechanical databases, provides much more information to support manufacturing, assembly, and test operations. The numbering used for the board detail drawing and/or the master pattern identifies the design database. Data files that are used for these processes should be created, identified, and controlled separately. By creating separate identities for these files, the data are accessed more easily because they do not have to be regenerated from the primary database. An example of common data that are generated and extracted from the CAE, CAD, and CAM tools is identified in Table 6.12.

TABLE 6.12

Data Typically Generated by CAE, CAD, and CAM Systems

Tool Type	Generated Data	Extracted Data
CAE	Circuit libraries	Linkage to physical library Component identification and electrical characteristics Circuit simulation data Schematic symbols
	Schematic/logic diagram	Electrical parts list Electrical net list Electrical testing data Schematic drawing
CAD	Layout	Linkage to circuit library Master drawing Board fabrication data Assembly data (part insertion) Testing data Tooling data (fabrication/assembly/test)
	Assembly documentation	PCA bill of material/parts list Assembly drawing Manufacturing/assembly/test instruction Documentation
	Support documentation	Detail drawings/associated parts list Specification drawings
CAM	Electrical Mechanical	Bare board electrical test data Gerber/ASCII/HPGL plot data Aperture data Drill/route data Panelization data

6.7 Archiving Data

As a general rule, data for completed designs should be removed from the design system to provide space for new designs. It is vital that these data are saved for future use with a well-managed archival process. This section describes naming conventions, media types and uses, and methods for tracking, controlling, and retrieving archived data.

6.7.1 Data Naming Conventions

Archived data normally are structured to include all the files required to completely describe a design. Placing the data files in various subdirectories organized as defined in Sec. 6.6.3 (circuit, mechanical, and layout data files) with the top subdirectory identified by the name of the PCA and its assembly number accomplishes this. The naming convention for each subdirectory identifies its function. Numbering and revision controls should coincide with the numbering structure identified within the PCA's documentation package.

6.7.2 Media Use

A company's size, the design tools that it uses, the size and complexity of the boards it designs, and its contractual requirements usually dictate the types of media used for archiving design data. File sizes are major factors requiring consideration during media selection. Some of the more common types of media used for archiving data are

- Diskettes
- Tape cassettes
- Reel-to-reel magnetic tapes
- CDs

Depending on the amount of data being archived, most companies will place multiple designs on the same medium to optimize storage. Normally, the data stored represent designs created with the same tool and are not mixed with data from other types of database structures.

6.7.3 Data Tracking, Control, and Retrieval

In order to minimize cost and schedule impacts, the archived data should be identified so that they can be located and retrieved easily. The basic identification scheme should include

- Medium type
- Part number
- Revision
- Archive date
- Archive location

Control and retrieval efforts can be minimized through the use of archive and retrieval request forms. The archive request form should have two sections. The first identifies the data files to be archived by name, number, and revision; the system used to generate them; and where they are located. The second section identifies who performed the archival activity, when it was completed, and where the archived data are located. The retrieval request form identifies the data to be copied from the archive and how and where the information is to be restored for use by the requester (medium, location, etc.).

6.7.4   Data Storage Processes and Controls

Most companies use a data-processing support organization for controlling computing systems, networking, and data integrity.

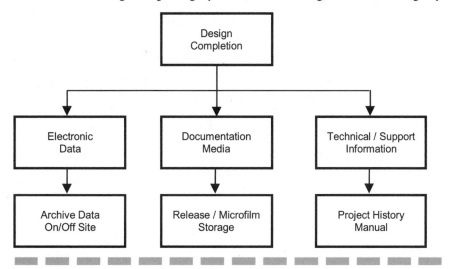

Figure 6.17
Media generation/control activities.

Archiving is normally performed by this function, which usually archives product data both internally and off-site for complete safety.

Besides electronic data storage, consideration also must be given to controlling drawings and technical and support information that is required for maintenance of product documentation. A data and configuration management function generally will do this. It provides the end product controls by structuring, assigning, and documenting the essential drawing and data identification requirements. Product changes are also identified and controlled. It ensures that proper documentation approval requirements are satisfied, copies distributed, and original drawings and data stored properly. Figure 6.17 defines media controls typically exercised following completion of a design. Table 6.13 identifies control activities and responsibilities.

TABLE 6.13

PCA Design and Support Data Management

Data Type	Contents	Control Activity	Responsibility
Electronic	Design data	Design organization/archive	Design/data processing
Documentation/drawings	Per-product requirements	Release/storage of originals and/or microfilm	Configuration/data management
Technical/support material	Concept through design product history	Storage of originals	Project manager

Design Revisions

There are a number of widely used industry standards that describe methods, guidelines, and basic procedures for identifying, documenting, and controlling printed circuit assembly (PCA) design revisions. These can be used to establish and manage cost-effective configuration change control processes that are applicable to all documentation associated with the design, procurement, manufacturing, assembly, and testing of PCAs. The key to structuring cost-effective controls for change implementation is understanding how different types of changes act on associated PCA documentation. Figure 3.16 shows a typical set of PCA design documents and common relationships and linkages.

Two simple forms can be created and used to control, identify, and track all PCA design and documentation revisions. There are two because one is intended to focus on changes that occur during the design process, whereas the other defines changes to be incorporated into documentation and data that have been

CHANGE PRIOR TO RELEASE

Project:		Date:	Page ___ of ___	Tracking Number:
Originator: _____ Ext: _____			Part Numbers Impacted: _____	
Requested By: _____ Ext: _____			_____	

DESCRIPTION OF CHANGES

Estimated Impact						Actual Impact					
Hours						Hours					
Review	Ckt Design	Layout	Check	Art	Total	Review	Ckt Design	Layout	Check	Art	Total

Estimated By:		Date:	Approved By:		Date:

Figure 7.1

Sample of form used to document changes made during PCA design.

released and placed under formal control. These forms should include the following types of information:

- Change identification (tracking number)
- Product identification (part number, description, and program)
- Origination date
- Originator (name)
- Change description (clear, concise, and legible—should show "was/now" or "from/to")
- Estimate (impact to project cost and schedule)
- Comments (ideas and suggestions for change implementation)
- Approval (name and date)

Figures 7.1 and 7.2 are examples of these two types of forms.

CHANGE FOLLOWING RELEASE

Project:		Date:	Page ___ of ___	Tracking Number:	
DESCRIPTION OF CHANGES					
Document Number	Document Title		New Rev	APPROVALS	DATE
				Originator:	
				Customer:	
				Engineering:	
				Manuf:	
				Test:	
REMARKS				AUTHORIZATION	DATE
				Proj. Engr:	
				Prog. Mgr:	
				Config. Mgmt:	
				Customer:	

Figure 7.2
Sample of form used to document changes made after PCA design is completed.

The following sections of this chapter will define methodologies and techniques used to control changes that affect schematics, layouts, bare boards, and assemblies.

7.1   Schematic and Parts List Revisions

The relationship of schematic and parts list to the overall PCA documentation and data package and the various types of information that they both provide should be understood completely. Throughout the electronics industry, a schematic and its electrical parts list are required to directly support "downstream" activities such as PCA layout, board fabrication, parts procurement, and assembly production and test. Figure 7.3 identifies the various types of data linkages commonly associated with a schematic/parts list database.

As shown in Fig. 7.3, if either the circuit parts list or the schematic diagram is revised, the database is affected. The circuit parts list identifies the following types of information:

- *Part type.* Resistors, capacitors, diodes, etc.

- *Source.* Supplier or manufacturer

- *Identification.* Complete part number

- *Electrical parameters.* Values, temperature ranges, etc.

- *Package.* Body style, number of pins, polarity indicator, etc.

- *Linkage data.* Internal identification, design libraries, circuit models, etc.

The circuit parts list is first used to verify that each component is described properly in the design library, which consists of two separate and interrelated elements:

1. *The circuit symbol library.* Created and maintained within a computer-aided engineering (CAE) tool suite, it contains a schematic symbol definition for each part type used in a design and its associated circuit properties.

2. *The physical part library.* Created and maintained within a computer-aided design (CAD) tool suite, it contains a

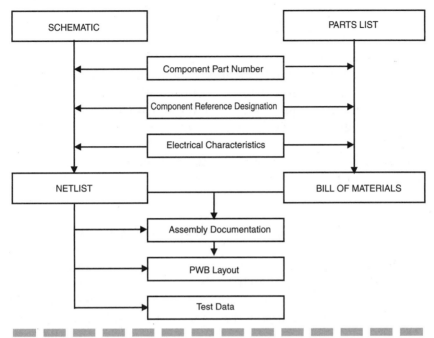

Figure 7.3
Linkages between schematic parts list and other PCA data items.

physical definition (part outline, termination locations, markings) for each part type used in a design.

If a parts list is changed, the revision may require changes to the data that affect the schematic and PCA layout, which directly use this information.

The circuit description provided by an electrical engineer in the form of a schematic and a parts list usually contains the following types of information:

- *Circuit symbols.* Used to represent parts used
- *Reference designations.* Used to identify circuit symbols/parts
- *Interconnections specified between schematic symbol terminations*
- *Signal identification.* Used to identify circuit interconnections
- *Standard notes.* Used to identify and control circuit design and layout requirements (grounding, termination of unused pins, etc.)

■ *Part types and quantities used,* including supplier number/identification

Schematic and parts list revisions normally are identified and controlled separately, although their information is linked together and affects the three primary areas of PCA design: circuit design, layout design, and support documentation. These relationships are shown in Table 7.1. The remainder of this section will provide information and guidelines for defining the controls and documentation impacts associated with changes to schematics and parts lists.

7.1.1 Identifying and Controlling Changes

Impacts on work in progress due to revisions should be identified on a form like the one shown in Fig. 7.1 until the schematic and parts list documents have been formally released and are under configuration control. Following the release and distribution of these documents, changes then can be identified and controlled using a form similar to that in Fig. 7.2. The typical types of revisions that can affect a parts list or schematic are

■ Changes in circuit connections

■ Changes in component numbers or part types

■ Changes in signal or net names

■ Changes identifying internal component elements (gates, amplifiers, or terminations)

■ Changes in unused pin assignments

■ Changes in external component or connector pinouts or termination identifiers

■ Design rule modifications (pull-up/decoupling components)

A major problem associated with the implementation of these changes is that because of schedule or budget constraints, many linked activities, such as the input to design libraries, creation of the schematic and parts list, circuit analysis, board layout, and materials procurement, are all being

TABLE 7.1

Impact of Schematic and Parts List Changes

Change	Circuit Design Impacts	Layout Impacts	Documentation Impacts
Part addition, deletion, type, physical configuration	Circuit symbol library linkages Circuit database Circuit performance simulation	Physical library linkages Component placement Routing Design rule check Thermal analysis	Bill of material Schematic drawing Master artwork Board detail drawing/data Assembly drawing/data Testing data Test tooling drawing/data
Part value/rating (same physical configuration)	Circuit symbol library linkages Circuit database Circuit performance simulation	Design rule check Thermal analysis	Bill of material Schematic drawing Assembly drawing/data Testing data
Signal name	Circuit database	No impact	Schematic drawing Testing data
Part number (same physical configuration)	Circuit symbol library linkages Circuit database	No impact	Bill of material Schematic drawing Assembly drawing/data Testing data

performed concurrently. Unless changes to any of these elements are synchronized properly with the others, errors could occur that may not be discovered until the PCA is being manufactured, causing costly hardware changes and major delivery schedule impacts. Typically, 60 percent of the changes to a PCA during or after production are caused by schematic or parts list changes that were implemented improperly during design. It is therefore extremely important that revisions to schematics or parts lists incorporated while a design is being created are carefully documented and controlled. This means that

1. Change notification methods should be defined, documented, and followed.

2. Changes always should be reviewed, evaluated for impact, and approved prior to implementation.

3. Correct incorporation of changes always should be verified (checked).

To help identify, control, and minimize changes affecting a parts list or schematic, the design review guidelines identified in Chap. 6 should be followed.

7.1.2 The Effect of Changes on a Design

Changing a parts list or schematic database may result in excessive cost and schedule impacts, depending on how and when the changes are implemented. The three primary areas of a PCA design described in Table 7.1 are used to identify the effects of changes to a parts list or schematic:

- Changes to a parts list during circuit design may affect
  - Design libraries
  - Schematic database
  - Circuit simulation/analysis
- Changes to a parts list during circuit board layout may affect
  - Part placement
  - Interconnection routing

- Marking
- Thermal analysis
- Changes to a parts list that occur during procurement may cause
  - Increased materials cost due to order cancellation, return and reorder, or unused material in inventory
  - Lengthened manufacturing schedule due to material shortages caused by delivery delays
  - Revision to procurement documentation
- Effects of changes to a schematic drawing that occur during circuit design include
  - Parts list revisions
  - Design library revisions
  - Rerun circuit simulation/analysis
- Effects of changes to a schematic that occur during circuit board layout include
  - Part placement revisions
  - Interconnection routing revisions
  - Rerun thermal analysis
- Effects of changes to a schematic that occur during circuit board procurement include
  - Increased materials cost due to order cancellation, return and reorder, or reworked or scrapped material
  - Lengthened manufacturing schedule due to material shortages caused by delivery delays
  - Revision to procurement documentation

Careful attention should be paid to the timing of implementation of changes that occur during the design activity. Until interconnection routing has started, circuit and part changes can be incorporated with relatively little impact on overall project cost and schedule, providing part procurement is not extensively affected. Once interconnection of a circuit board is underway, any revision that requires substantial changes to placement or routing should be analyzed carefully. Consideration can be given to possible alternatives such as incorporating it at the next version of the product or modifying the hardware as part of the production process.

7.1.3   Interchangeability Considerations

Following a PCA's initial or prototype build, the design may require circuit enhancements or part changes prior to production start-up. Both parts list and schematic changes must be identified and controlled in order to maintain interchangeability between the existing and new versions of the design. Changes that do not affect the product's overall form, fit, or function can be implemented and controlled by incrementing the revision letters of the affected drawings.

The majority of changes, however, usually affect an assembly's form, fit, or function. This results in having to change the PCA's part number, since the original and new versions are no longer interchangeable. The simplest and most economical way to re-identify a design configuration is by retaining its root part number and changing its dash number. Changes can be identified properly while retaining the majority of the original documentation and only adding the information describing the modifications to the original design.

A major concern that also should be addressed is the linkage between the documentation and all the electronic data associated with a product. Usually, when a change has been implemented, the electronic data supporting the previous configuration of the product are no longer correct. Unless those data are also updated and reidentified to properly reflect the revised design configuration, the possibility exists for using an incorrect version for fabricating, assembling, or testing the PCA.

7.1.4   Change Documentation and Data Control

The two forms previously reviewed in Figs. 7.1 and 7.2 can be used to control change documentation activities for both in-process and released documentation. The dash-number form can readily distinguish, identify, and control changes that affect a product's interchangeability. The parts list and schematic documentation can be controlled by the use of revisions, although the schematic documentation, especially if it is linked to other databases, requires some additional controls to support and maintain accurate data extraction. Figure 7.4 describes how changes can be controlled within the note section of a schematic, with Fig. 7.5 showing change controls within the body of the schematic.

Data management for parts list and schematic documentation should be flexible enough to support a variety of product manufacturing conditions and situations. Some of the issues that should be reviewed prior to structuring data controls are

■ Configuration control requirements

■ Costs associated with the implementation plan

■ Controls and methods for change implementation

■ Documentation controls and requirements

■ Data controls and requirements

■ Data and documentation linkages

■ Data and documentation archiving controls and requirements

Both ASME Y14.35M and MIL-STD-100, which are standards most commonly used to structure and identify configuration controls, consider these issues.

Changes that are controlled within the note section of a schematic drawing always should start after the last of the standard notes. In the following example, changes start with note number 6. Figure 7.5 shows the change implemented in the body of the schematic.

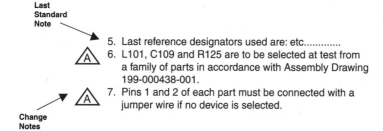

Figure 7.4
Example of schematic changes described by notes on the drawing.

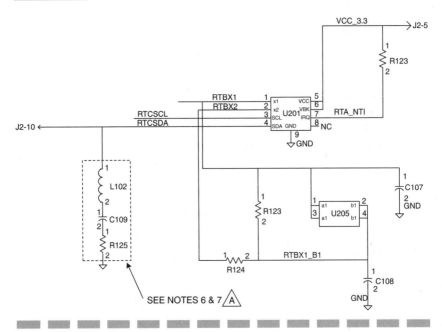

Figure 7.5
Changes implemented in the body of a schematic.

7.1.5 Hardware Configuration Control

In order to effectively ensure that the configuration integrity of a product is maintained, data controls should be established and used as a PCA design evolves toward maturity. The primary information identified in hardware configuration control structuring is

■ Document type and format standards to be used

■ Part number and release standards

■ Drawing and part title standards

■ Review and approval requirements

■ Data and documentation linkages

Hardware drawing release and change history can be used as a data source for evaluating a product's current status and can provide valuable information for making cost-effective decisions,

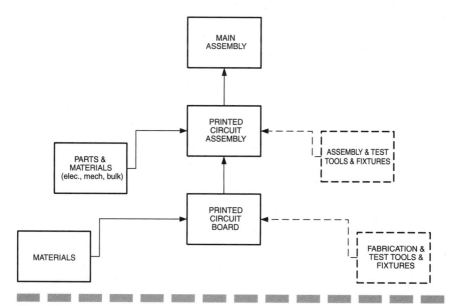

Figure 7.6
Product family tree.

such as when to incorporate changes or when to produce the next lot of boards.

A family tree is the most common method for identifying and controlling hardware relationships. Figure 7.6 presents a generic example of a family tree for a product.

7.2 Layout Changes

Design changes may occur while an initial layout is being developed or after the layout has been completed and the PCA manufacturing documentation and data have been released for production. The most common areas of a layout affected by design changes include

- Physical/functional part data
- Part placement
- Circuit interconnections

- Material selection
- Board construction
- Board physical features

Although layout changes may affect interrelated data and documentation, they can be implemented cost-effectively, provided the layout change process is done in a controlled way. In order to achieve such control, the major impacts of the changes must be identified. This section identifies some of the more common methods and techniques used to control and document layout changes. The effect that design changes have on a layout, interchangeability considerations, and implementation decisions (revision versus relayout) will be reviewed.

7.2.1 Identifying and Controlling Changes

Types of design changes that may occur during the initial layout include

- Part additions, deletions, revisions
- Circuit modifications
- Part placement changes
- Interconnection routing changes

If identified early, incorporation of design changes into a layout can be done without significant impact on nonrecurring cost or completion schedule, since the impact on related data is minimal. Once a layout is completed and manufacturing documentation and data are issued, the effect of design changes on cost and schedule can increase dramatically. At this point, a tradeoff analysis should be done, comparing the cost and schedule impact of revising the original layout and associated documentation versus modifying the hardware itself to incorporate the required changes. Typical changes that may be done during production of hardware following documentation release include

- Interconnection modifications (cuts, jumpers, drillouts, etc.)
- Part replacement, additions

- Physical board or assembly modifications
- Marking revisions

It should be understood that these modifications also must be documented fully in order to ensure that they are done in a controlled and repeatable manner, so the cost associated with this activity must be considered when making the tradeoff. The most common methods and techniques used for controlling changes are as follows:

- Changes during the initial layout:
  - Check plots (changes identified by color coding design plots)
  - Control forms (as identified in Fig. 7.1)
  - Online CAD system verification (verify change difference reports)
- Controlling changes following documentation release:
  - ECN (engineering change notice)
  - Drawing revisions (letter controls)
  - Part number modification (dash-number control)

7.2.2   The Effect of Changes on a Design

As changes occur, the effects on the original design should be reviewed carefully. Some changes may require a total redesign effort that significantly affects a product's production cost and delivery schedule. Regardless of the type of change to be implemented, they all affect the design with various penalties.

As changes are identified, they should be reviewed in detail by all affected functions prior to implementation. Some of the activities that may be affected by changes are

- Simulation and analysis (circuit, thermal, mechanical)
- Generation of manufacturing and reference documentation
- Material and component procurement
- Board fabrication
- Assembly
- Testing

Understanding the effects of changes on all the associated data and documentation will help to determine the most cost-effective way to plan and implement a design change while controlling the product's integrity. Controls and methods should be in place to aid in determining the best time in a design cycle to incorporate changes.

7.2.3 Interchangeability Considerations

As a PCA evolves, changes that occur further downstream in the development process tend to have a greater effect on its overall interchangeability. Interchangeability of a PCA relates to its ability to replace another assembly without changing the performance of its home system or require readjustment of that system. The following major factors should be considered when evaluating the effect of a design's change on a PCA's interchangeability with configurations that already exist:

- Physical form and fit
- Functional performance
- Test procedures
- Predicted reliability
- Field maintenance, repair, and adjustment
- Identification and reference documentation (drawings, manuals, etc.)

If a PCA undergoes extensive changes that affect any of the preceding factors, its documentation should identify that the configuration is different from the preceding version, even if it performs the same function and fits in the same place. In order to control product integrity, some or all of the following actions may be required when implementing a configuration change:

- Assembly reidentification
- Assembly requalification (test and integration)
- Release of a new documentation package and manufacturing data

During a PCA's design cycle, interchangeability issues only arise following the initial release of its documentation. The majority of layout changes that occur usually are driven by bare board or assembly modifications that are required to resolve manufacturing or performance problems. Hardware modifications commonly made during the fabrication of bare boards or assemblies should be identified and controlled within the assembly's documentation.

7.2.4   Change Documentation and Data Control

If properly structured and maintained, documentation and data control processes can easily allow cost-effective change implementation. Change control is very important during the initial PCA design phase, primarily due to data linkages and activities involving procurement of long-lead parts and materials. Two common industry standard methods are used for controlling changes. One is to use revision letter version controls for the documentation, and the other is to reidentify (renumber) all the documents and reidentify the hardware they define.

Advancing the revision letter of drawings following documentation release usually identifies and controls those design changes which do not affect form, fit, or function of a PCA. Some of the more common changes of this type are

- Board markings
- Reference designators
- Notes
- Drafting errors or clarifications

Reidentification of the documentation usually is required if the circuit board or PCA has already been built and/or delivered. This may seem time-consuming and costly, but by using currently available CAD tools, a design database can easily be copied, edited, and reidentified.

The more common types of layout changes that require a part number change are

- Physical dimensions (*XYZ*)
- Part additions, deletions, modifications
- Circuit revisions
- Use of different materials or construction

7.3   Bare Board Modifications

How bare board modifications are implemented and documented may vary depending on program or customer requirements.
Figure 7.7 identifies before and after conditions following incorporation of some typical bare board modifications.

7.3.1   Effects of Bare Board Changes on Other PCA Data and Documentation

The types of data and documentation commonly affected by
bare board changes are identified in Fig. 7.8. The following specific areas within them should be reviewed to identify specific
change impacts on data and documentation:

- Drawings
  - Fabrication instructions
  - Detail views
  - Assembly information
  - Schematic
  - Design layout
  - Artwork for stencils/screens
  - Test tools and fixtures

- Electronic data
  - Net list
  - Bill of material
  - Drill file
  - Part placement file
  - Artwork plot files
  - Test files
  - Board machining/profiling files

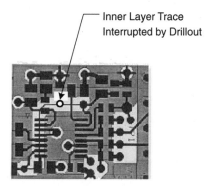

Inner Layer Trace
Interrupted by Drillout

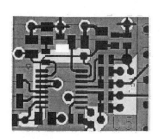

Top View of Multilayer Board Showing Inner Layers

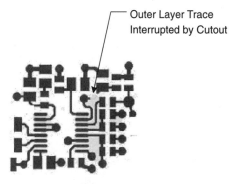

Outer Layer Trace
Interrupted by Cutout

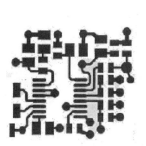

Top View of Multilayer Board Showing Outer Layer

Figure 7.7
Before and after modification views of a multilayer board.

7.3.2 Processes, Methods, and Documentation

It is important to establish a set of standardized processes for making and documenting each of the various types of modifications usually performed on bare boards. Again, these may vary depending on program or customer requirements, but these methods should be established and controlled to ensure the consistency and quality of the end results. For each of the different types of modifications, the documented standard process should specify

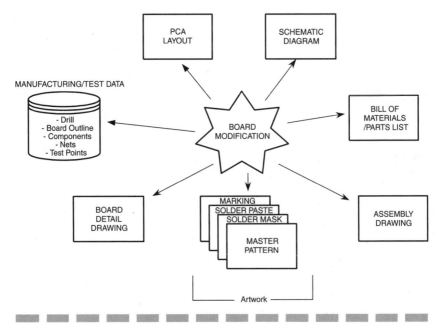

Figure 7.8
Documentation/data potentially affected by bare board modification.

- The type of change
- How the data and documentation are to be revised
- Alternate methods to be used to make the change
- Quality requirements

Figure 7.9 shows how some typical board modification processes are implemented and documented.

7.3.3 Documentation and Data Control

Table 7.2 identifies the major types of documentation usually associated with bare board modifications and their data linkages. When establishing the documentation and data modification process, consideration should be given to the following factors:

- Streamlining to minimize the amount of documentation required

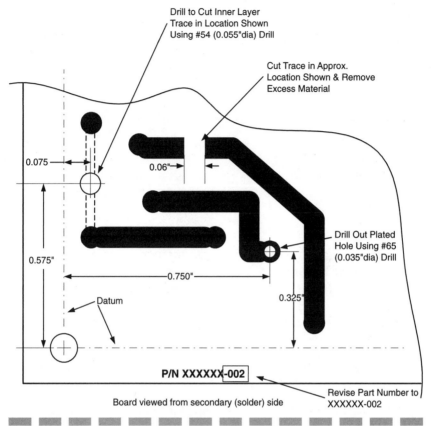

Figure 7.9
Typical bare board modifications.

- Provision of a sufficiently detailed description of exactly how to make the change
- Definition of the quality requirements levied on the end result
- How to reidentify a modified board

7.3.4 Marking of Modified Boards

Board marking methods and techniques should be well defined as part of the modification process and indicated on the revised

TABLE 7.2

Documentation Usually Used for Implementing a Board Modification

Documentation	Data Linkage
Change authorization (such as engineering change notice)	Identifies/documents/controls change and hardware reidentification
Fabrication drawing	Specifies how the physical modification is implemented (see Fig. 7.9).
Special work instructions	Describes manufacturing/test methods and procedures for accomplishing the modification

documentation. A bag and/or tag method frequently is used for modified bare boards that do not have sufficient area for direct marking. If a board is too small to attach a tag to it, a bag large enough to enclose the physical board and its identification tag should be provided. The identification tag should include such information as

- Board part number
- Revision designator
- Manufacturing lot number and/or board serial number
- Date of modification
- Controlling change document number(s)

The tag (or the information on the tag) should accompany the board through all subsequent assembly, test, and integration processes and made available, if needed, for future reference.

Hand stamping is commonly used throughout industry as a preferred method for marking modified boards. The same information required for the tag should be marked on the board. In addition, concern for circuit performance and legibility and permanence makes it important that the board designer specify

- Size (height, width, and thickness) of characters
- Marking location
- Ink type

If space permits, labels also may be used to mark modified boards. In addition to legibility issues, the designer should ensure that the label will survive all subsequent manufacturing and test processes, as well as the PCA's expected operating environment, especially as relates to adhesion of the label to the board. The label also should be located where there is no chance of entrapment of corrosive materials underneath it.

7.4 Assembly Modifications

PCA designers should understand the impact of various types of modifications to existing assemblies or existing data and documentation. Some of the common causes of assembly modifications are

- Part substitution
- Circuit performance shortfall
- Change in performance requirements
- Change in customer/program requirements
- Manufacturing or test problems
- Producibility improvements
- Technology improvements

This section identifies and provides an understanding of some of the common processes, markings, documentation, and controls used for assembly modifications.

7.4.1 Effects of Assembly Modifications on PCA Data and Documentation

The types of data and documentation commonly affected by changes are identified in Fig. 7.10. The following specific areas within them should be reviewed to identify specific change impacts on data and documentation:

- Drawings
  - Assembly instructions

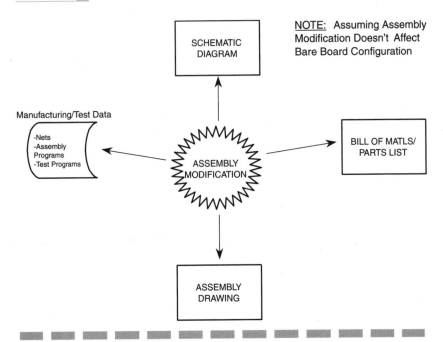

Figure 7.10
Potential documentation/data affected by assembly modification.

- Detail views
- Testing information
- Schematic
- Assembly tools and fixture
- Test tools and fixtures
- Electronic data
  - Net list
  - Bill of material
  - Part placement file
  - Test files

7.4.2 Processes, Methods, and Documentation

As was the case for modification of bare boards, it is important to establish a set of standardized processes for making and docu-

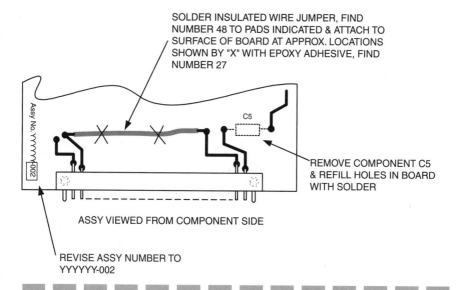

SOLDER INSULATED WIRE JUMPER, FIND
NUMBER 48 TO PADS INDICATED & ATTACH TO
SURFACE OF BOARD AT APPROX. LOCATIONS
SHOWN BY "X" WITH EPOXY ADHESIVE, FIND
NUMBER 27

C5

Assy No. YYYYY-002

REMOVE COMPONENT C5
& REFILL HOLES IN BOARD
WITH SOLDER

ASSY VIEWED FROM COMPONENT SIDE

REVISE ASSY NUMBER TO
YYYYYY-002

Figure 7.11
Typical assembly-level modifications.

menting each of the various types of modifications usually performed on assemblies. Again, the modification methods and procedures used may vary depending on program or customer requirements. For each of the different types of modifications, the documented standard process should specify

- The type of change
- How the data and documentation are to be revised
- Alternate methods to be used to make the change
- Quality requirements

Figure 7.11 shows how some typical assembly modification processes are implemented and documented.

7.4.3 Documentation and Data Control

Table 7.3 identifies the major types of documentation usually associated with assembly modifications and their data linkages.

TABLE 7.3

Documentation Usually Used for Implementing an Assembly Modification

Documentation	Data Linkage
Change authorization (such as engineering change notice)	Identifies/documents/controls change and hardware reidentification
Assembly drawing	Specifies how the physical modification is implemented (see Fig. 7.11); board modifications, if required are also documented here
	Identifies and documents changes to PCA bill of materials/parts list to accommodate part addition/deletions, jumpers, and additional material/process requirements
Special work instructions	Describes manufacturing/test methods and procedures for accomplishing the modification

When establishing the documentation and data modification process, consideration should be given to the same factors described earlier in Sec. 7.3.3.

7.4.4 Marking of Modified Assemblies

Use the same procedures described for marking of modified boards (see Sec. 7.3.4).

Design Organization and Management

A printed circuit assembly (PCA) design organization's structure and the guidance and supervision of its activities have a significant impact on how effectively its efforts will satisfy cost, schedule, and quality performance requirements. Regardless of the size of a company, if PCAs are designed internally for its products, an identifiable function should be established that is responsible for performing all related activities.

Design of PCAs requires the use of a unique set of methods, processes, and related tools to effectively accomplish the following activities:

- Circuit board layout
- Design library management
- Documentation and data generation
- Checking and verification
- Tool and equipment support
- Documentation and maintenance of operating procedures
- Training
- Interfacing with other functions

These tasks may all be accomplished by one person or by a very large group of people. In either case, an organizational structure and associated operational procedures that support that structure are needed to ensure design consistency and quality.

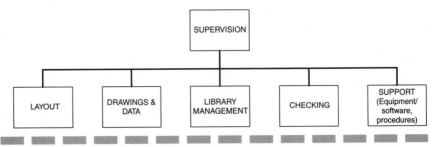

Figure 8.1
Typical PCA design organization structure.

8.1 Structure of the Design Organization

Companies are configured in many different ways, some managing their engineering and development efforts as independent programs and some performing them within functional organizations. Most, however, have a centralized circuit board design operation that services all program requirements.

The primary features of a typical organization are derived from the key activities and tasks that are associated with developing and documenting a PCA design. The basic processes used to produce a design are standard throughout the electronics industry and vary only in the details. The following identifiable functions usually are found in a PCA design organization:

- Circuit board layout
- Management of libraries
- Documentation and data
- Checking
- Tool and equipment support
- Processes and procedures
- Supervision

Figure 8.1 shows the structure for a typical PCA design organization. Although other important activities are performed, such as training and interfacing with functions outside the organization, these are usually carried out within one or more of the preceding areas. Large companies sometimes tend to fragment and disperse a PCA design organization by assigning some of these activities to external functions such as drafting for documentation or information systems for computer tool and equipment support. Unfortunately, this organizational approach tends to create significant inefficiencies of overall operation due mainly to the unique technical and schedule constraints associated with designing PCAs, the close interrelationship among all the activities, and dealing with priority conflicts within the other functions.

8.1.1 Layout Function

The core of a PCA design organization is its layout group. It is responsible for performing all design-related activities and producing a layout that satisfies all technical requirements. Since layout constitutes the bulk of the work done, it has the greatest effect on the ability of the organization to meet its nonrecurring cost of design and schedule commitments.

The layout function is the main interface with circuit designers, mechanical engineers, and the manufacturing, assembly, and test organizations and is responsible for correctly identifying and interpreting all technical requirements. Processes and procedures that define how layouts are to be developed are unique to each PCA design organization, depending on what types of assemblies are required, their complexity and performance requirements, and the types of tools used to produce the layouts. When establishing these processes, methods for revising existing layouts also should be included.

8.1.2 Library Management

The circuit symbol and physical component libraries form the foundation of a PCA design, and the creation, maintenance, and control of these design libraries are critical functions. A librarian is usually designated and has the following responsibilities:

■ Defining the type and structure of the data required to be entered in each of these libraries and how they are linked to each other (Figure 8.2 contains examples of typical schematic symbol and physical component library entries.)

■ Documenting and verifying the data to be entered (including interfacing with component suppliers, when required)

■ Establishing component identification (naming, part number) conventions

PHYSICAL / LOGIC LIBRARY ENTRIES

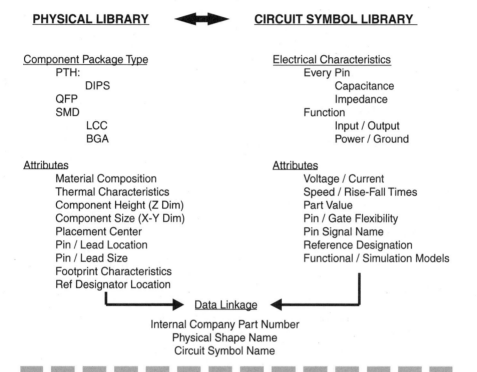

PHYSICAL LIBRARY ◄──► **CIRCUIT SYMBOL LIBRARY**

Component Package Type
    PTH:
        DIPS
    QFP
    SMD
        LCC
        BGA

Electrical Characteristics
    Every Pin
        Capacitance
        Impedance
    Function
        Input / Output
        Power / Ground

Attributes
    Material Composition
    Thermal Characteristics
    Component Height (Z Dim)
    Component Size (X-Y Dim)
    Placement Center
    Pin / Lead Location
    Pin / Lead Size
    Footprint Characteristics
    Ref Designator Location

Attributes
    Voltage / Current
    Speed / Rise-Fall Times
    Part Value
    Pin / Gate Flexibility
    Pin Signal Name
    Reference Designation
    Functional / Simulation Models

Data Linkage
    Internal Company Part Number
    Physical Shape Name
    Circuit Symbol Name

Figure 8.2
Typical physical and schematic symbol library entries and their relationships.

- Identifying related data required (design rules, land patterns, test points, circuit function intelligence, etc.)
- Inputting data to the libraries and verifying its correctness

The library management function also should perform scheduled "housekeeping" activities to keep libraries at a manageable size by eliminating obsolete and duplicate component entries. In companies that have an established components engineering organization, the PCA librarian may be required (or desire) to obtain its concurrence or approval before entering a new component or revising an existing entry.

8.1.3 Documentation and Data

There is usually a group within the PCA design organization that has responsibility for preparation, distribution, and control of all related drawings and manufacturing data. This material may be prepared using a variety of methods and processes, depending on the source of the data.

If the layout was created manually, data and drawings are prepared manually unless the design information is captured in computer format by digitizing the layout. Data and drawings are usually generated using computer-aided techniques if the layout was produced using a computer-aided design (CAD) system. Many printed circuit CAD systems include capabilities for producing manufacturing drawings, and all systems have the ability to produce data files needed to support board fabrication, assembly, and test operations. Using a CAD system to directly generate drawings is the most effective approach. However, if these resources are limited and are needed to produce layouts, then preparing drawings may have to be treated as a lower-priority activity, and the drawings may be produced by transferring the design data to a computer-aided drafting system or by drafting them manually.

The documentation function also usually has responsibility for conducting drawing reviews, obtaining drawing approval signatures, and releasing them into a document and data control system for reproduction, distribution, and storage. This function also incorporates drawing revisions, as required. The procedures used for documentation, release, and revision activities usually are based on preexisting company drafting standards and are modified only if required to reflect requirements that are unique to the PCA environment. Data files are identified and archived using a company's standard procedure for storing software files.

8.1.4 Checking Function

The checking function verifies correctness of a design and its associated documentation and data files. It is usually prudent to have a separate group perform this activity or at least have it done by someone other than the one who created what is to be checked.

Checking activities should be performed using well-defined procedures, and sufficient schedule time and budget should be allocated for it to be accomplished properly. Checking methods will vary, based on whether a PCA design and documentation were created manually or by using computer-aided tools. Regardless of how a layout was produced, however, there are common elements of a design that should be verified. These are

- Comparing the layout with the schematic data for correct use of components and accuracy of their interconnections
- Ensuring adherence to design rules and requirements
- Assessing the producibility of the design
- Checking drawings and data files to verify that they correctly and completely define the design to be built and tested and that they are prepared in accordance with all appropriate drafting and data standards

The checking effort should not be viewed as a "necessary evil" and treated casually. Errors that are identified during the design and documentation cycle usually can be corrected easily and inexpensively. Once hardware is built, correcting these same errors can have a significantly greater impact on overall program costs and completion schedules.

8.1.5 Support of Design Tools and Equipment

Tools, equipment, and software should be maintained properly and updated as needed to keep a PCA design organization functioning effectively and efficiently. This is applicable for operations that produce designs manually but is particularly critical for those which use computer-based methods.

Support activities should deal with maintaining the proper functionality of the software and equipment used by design and documentation personnel in their day-to-day operation of the systems. The support function also should be responsible for performing routine "housekeeping" activities such as data backup and archiving.

It is important that standard support practices be established, documented, and followed to ensure that system availability is maximized. Documented procedures should include both standard and emergency instructions for

- Equipment use (power-up, power-down, software initialization, program lockups, etc.)
- Identification and resolution of hardware and software problems
- Preventive maintenance
- Interfacing with suppliers' technical support organizations
- Record keeping to verify proper performance of maintenance and problem resolution activities
- Incorporating and validating equipment and software modifications and updates
- Establishment and maintenance of a historical description of all system hardware and software elements, starting with the baseline configuration and including all modifications and updates

Many, if not all, elements of a design system usually are covered by a maintenance agreement, and it is the responsibility of the in-house support function to interface with these suppliers' support organizations. This activity includes notifying them of problems, working with them to resolve outstanding problems, and verifying that the problems have been closed out properly.

Other support activities may include monitoring environmental conditions that may detrimentally affect system operation (temperature, humidity, power source, communication links, etc.), managing data storage and retrieval, generating and documenting custom software programs, routines, scripts, and patches, and maintaining a supply of needed consumable materials and supplies.

8.1.6 Development, Documentation, and Maintenance of Processes and Procedures

Regardless of the size of a PCA design organization, standard operating procedures should be documented and followed. Using standard procedures will help maximize efficiency and aid in the development of quality designs by establishing a uniform, disciplined methodology. Some of the advantages of this are

- Personnel have a well-defined process "roadmap" to follow and do not have to spend time developing their own procedures.
- Performance data can be gathered from the implementation of standard processes and used to improve them.
- Accurate budget and schedule estimates can be produced based on the consistency of past history.
- Training is more effective, and new technical personnel become productive quickly.
- Problems can be identified and resolved easily.
- Standard processes can be tailored to unique job requirements.
- The impact of personnel turnover is minimized.
- Processes can be revised to accommodate use of new tools and equipment.

Responsibility for developing, documenting, and maintaining standard processes should be assigned to specific personnel in the PCA design organization. Most of the inputs to these processes will come from the people who are directly creating the layouts and the documentation, as well as from the people who are supporting those activities. Identification and capture of process metric data can be accomplished by using work logs, which can be analyzed and the data used to optimize the processes. At minimum, standard processes should be put in place for

- Layout
- Library management
- Documentation and data
- Checking
- Tool and equipment support
- Training
- External interfaces
- Process improvement

8.1.7   Training

It should not be the intent of internal training to teach people how to design or document PCAs; they should already have such skills. The basic purpose of the training is to provide and maintain knowledge of technical requirements associated with company-unique products, processes, tools, equipment, and organizational structure. It should be provided to all supervisory, design, documentation, and support personnel who are directly involved in the design of PCAs. Another important objective is cross-training of people in the organization to enable them to have the knowledge needed to accomplish a variety of assignments.

Although on-the-job learning is very important, it should be supplemented with some formal training, especially to provide familiarization with any standard operating procedures that are in place. Also, many suppliers of computer-based design and documentation systems furnish marginally effective training for users of their software or associated equipment, so this type of training may have to be developed and provided internally. As standard operating processes are revised to improve them or to accommodate new requirements, some retraining probably will be required.

8.1.8   Supervision

How PCA design and documentation activities are supervised is very dependent on the way the organization is structured. Assuming that a centralized circuit board design function exists that services all program requirements, the supervisor of that resource is responsible for producing a PCA design that meets all performance requirements, is producible, and is completed within specified budget and schedule constraints. This means managing the personnel and the tools and equipment used to accomplish these activities. To do this requires extensive technical knowledge of PCA design, manufacturing, assembly, and test techniques and requirements, as well as an in-depth understanding of the workings of the specific design and documentation tools and equipment being used.

A supervisor's primary areas of responsibility generally include

- Establishment and maintenance of a viable PCA design and documentation resource that is responsive to the needs of the programs being supported
- Preparation of job cost and schedule estimates
- Reporting of the budget, schedule, and quality status of all work in progress
- Selection and assignment of personnel and assessment of their performance
- Verification that all in-place processes and procedures are being implemented properly
- Assessment of the capability and capacity of personnel and their tools to meet present and future needs of the programs and establishment and implementation of plans to upgrade the resources where needed
- Interface with all functions and organizations that are involved in fabrication, assembly, test, and use of the PCAs that have been or are being designed

8.2   External Interfaces

In the course of developing a PCA design and its associated documentation, personnel in the organization performing the work usually have to deal effectively with a number of different external functions. It is important that standard procedures be established for managing these linkages in a consistent and efficient manner.

Figure 8.3 shows that PCA-related activities involve the flow of technical and administrative information to and from a variety of areas, and the success of the design and documentation effort is directly related to how effectively and efficiently these interorganizational interfaces are implemented.

Key organizational links that usually need to be managed include

- Program/project management
- Systems engineering
- Circuit design

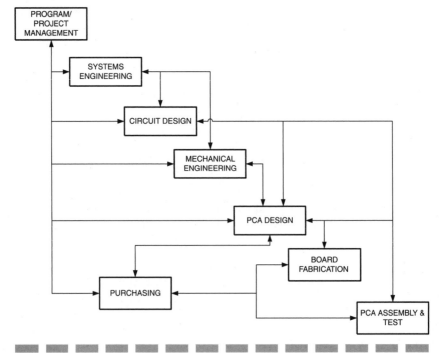

Figure 8.3
Typical flow of technical and administrative data.

- Mechanical engineering
- Material procurement
- Printed circuit board fabrication
- Production assembly and test
- Purchasing

8.2.1 Program Management

There is usually a management organization within a company that defines and directs the administrative aspects of technical activities on a program. It is generally responsible for establishing a program's schedule requirements, allocating budgets for technical efforts, and defining high-level technical requirements.

The PCA design organization's supervisory function is usually the primary interface with management on all "business-related" activities. Dealings with such areas as marketing and internal business organizations or external customers usually are coordinated with and accomplished through program management.

Detailed schedules and budgets for specific design and documentation tasks are developed and submitted by the PCA supervisor based on the scope of task requirements defined by program management and availability of resources. Program management usually has the responsibility for approving schedules and providing funding for the tasks. A standard methodology should be established and used for preparing cost and schedule estimates to provide the credibility and data backup needed to adequately justify them.

PCA supervision is usually required to regularly report on the status of the progress of authorized design and documentation activities relative to their established work schedules and allocated budgets. Any problems that may threaten or cause a budget overrun or schedule slip should be described in the status reports, along with a description of actions being taken to mitigate them. Here again, a standard progress assessment and reporting process should be established.

8.2.2   Circuit Design and Mechanical Engineering

Circuit design and mechanical engineering functions usually define the detailed technical requirements for a PCA. How effectively the processes associated with these interfaces are implemented has a significant effect on the success of the overall PCA design and documentation effort. If procedures defining external interfaces are specified nowhere else, they should be formalized and documented for the flow of information between PCA design and the circuit and mechanical engineering functions. The key elements of such a process description should include

■ The format and structure of the data initially provided to the PCA design function, their timing, and the mechanisms used for transmission

- The methodology for revising technical inputs
- Interaction between designers and engineers during work in process
- Organization and performance of design reviews
- Layout approval authority
- Documentation review and signoff

8.2.3 Board Fabricators

One of a PCA design function's primary responsibilities is provision of information required for bare board fabrication. This may consist of either a set of manufacturing drawings and phototools, computer-generated data files, or a combination of both. Manufacturing data supplied for in-house fabrication usually are transmitted directly to the board shop, whereas material going to an external fabricator frequently is controlled by the company's procurement organization, which provides the manufacturer with a purchase order and then authorizes the PCA design organization to transmit the required manufacturing data electronically. If actual drawings or phototools must be sent, they are usually forwarded through the procurement organization.

Once a board producer has received and evaluated the manufacturing data, there may be a need to interface with the PCA designer to discuss and interpret some of the specified requirements, to resolve possible problems associated with meeting them, or to suggest modifications that would improve board producibility or quality. It is important that standard processes be established defining the content and structure of manufacturing data, how they are to be identified and controlled, and how they are to be transmitted.

It is equally important to control the technical interface between a board manufacturer and the PCA designer. Any discussions or agreements between them that may involve a revision in requirements should be documented, and all potential cost or schedule impacts should be identified and evaluated before a change is authorized by responsible supervision. When dealing with an external fabricator, changes should be authorized through the procuring organization, since they may impact preexisting contractual agreements.

The process for documenting agreed-to revisions to board features or requirements also should be defined. Issues regarding how drawing and data file changes are to be made should be addressed, including authorizing drawing markups (redlining), reissuing data files, revising and rereleasing drawings prior to material acceptance, and others.

8.2.4 Assembly and Testing Functions

Procedures similar to those followed in dealing with bare board fabricators should be put in place for managing interfaces with PCA assembly and test operations. They also should address issues concerning how data are structured and transmitted and the documentation and authorization of requirements interpretations and revisions.

8.2.5 Design Tool Support Activities

A very close working relationship should be established with the organizations that support and maintain the software tools and equipment that are used to perform PCA design and documentation activities. These are critical resources, and their availability must be maximized to meet budget and schedule commitments.

Ideally, support activities should be performed by personnel within the PCA design function, as suggested in Sec. 8.1.5; however, some companies choose to assign some or all of that responsibility to another organization, such as data processing or information systems. If the latter approach is used, PCA supervision should ensure that a documented procedure is in place defining support methodology and responsibilities and that it is agreed to by both organizations.

The process should cover all the areas of activities described in Sec. 8.1.5. In addition, since quick response to identified problems is key to minimizing design system downtime, the process should define

■ How problems are reported and documented

■ Expected response time to problems (normal and emergency)

■ Multishift coverage, if appropriate

■ How problems are resolved and closed out

It is important to obtain assurance that qualified personnel are assigned, that they are familiar with the software and equipment being used, that a procedure for obtaining supplier help is in place, and that appropriate support is available when needed.

8.2.6 Company Purchasing Function

The need frequently arises for PCA design personnel to deal with external suppliers concerning technical questions and issues. Typically, such discussions may be with

■ Component and part suppliers

■ Material suppliers

■ Bare board fabricators

■ Assembly and test operations

■ Manufacturing tool and fixture fabricators

■ Suppliers of design and documentation software tools

All contacts with representatives of companies that are supplying products or services or negotiating to supply them should be coordinated with the in-house purchasing function. A procedure to ensure this should be in place within the PCA organization.

APPENDIX A

GLOSSARY

(Definitions provided are specifically applicable to the design, fabrication, assembly, and test of printed circuit assemblies.)

A/D converter   A functional circuit element or component that converts analog signals to digital signals.

active component   An electronic component whose parametric characteristics change while operating on an applied signal.

additive process   A process for fabricating a printed circuit board by selectively depositing conductive material to define a conductor pattern on unclad, chemically treated laminate material.

adhesive characteristics   Adhesive and cohesive strength over its expected operational environment are the essential characteristics of an adhesive material. Adhesion is the strength of the bond at the interface surface. Cohesion is the bulk (cross-sectional) strength of the adhesive material.

adhesive dam   A method of controlling the flow of adhesive material on inner layers of a multilayer board during lamination by providing an etched border of copper around the outer edges of the conductor layers.

aging   Mechanisms that cause degradation of properties or performance over a period of time.

American National Standards Institute (ANSI)   An organization that establishes national and international standards. ANSI publications that are most applicable to printed circuits are those that define drafting procedures and documentation formats.

American Society for Testing and Materials (ASTM)   An organization that provides a forum for the development and publication of test standards and procedures for materials, products, systems, and services.

analog   A type of circuit that deals with continuously varying voltage or current values that represent physical quantities.

annular ring   The portion of conductive material (usually a pad) that completely surrounds a hole.

aperture   A predetermined shape (round, square, oblong, etc.), size (width, diameter), and type (draw or flash) that is exposed on artwork film by a photoplotter.

aperture library   A collection of standard aperture descriptions.

aperture wheel   Contains a set of individual physical apertures (or software definitions of apertures) that are specific for plotting a type or family of circuit board artworks.

application-specific integrated circuits (ASICs)   An electronic device that is custom designed for a particular application. ASICs may be designed by interconnecting existing circuit subcircuits (functional circuit building blocks that already exist in a library) or by developing and connecting a completely new set of circuit functions.

archiving data   The process of relocating all the design and data files required to produce an end product into a retrievable storage area.

artwork   An accurately scaled photoprocessing tool that is used to fabricate a printed circuit board.

ASCII format   A code for representing English characters where each character is assigned a number from 0 to 127. Text in ASCII code is easily transferred from one computer to another.

aspect ratio   The ratio of length or depth of a hole to its preplated diameter.

assembly   A number of parts or subassemblies or any combination thereof that are joined together.

assembly drawing   A document that depicts the physical relationship of a combination of parts and subassemblies that form a higher order assembly.

atmospheric variations   Changes in operational or storage conditions.

automated assembly   Automatic component placement and attachment to a printed circuit board.

automated optical inspection (AOI)   A vision system that captures and stores an image and compares it to an expected image and/or a set of design rules to detect errors on printed circuit artwork, boards, or assemblies.

automated part placement equipment Electronically controlled equipment used for assembling components on circuit boards.

automated test equipment Electronically controlled equipment utilized for various testing operations to verify end-product integrity.

autoplacement The activity by CAD software that automatically places components on a circuit board layout based on preset rules.

autorouter CAD layout software that automatically determines the placement of interconnections on a circuit board based upon predetermined design rules.

axial lead Wire termination extending from a component or module body along its longitudinal axis.

B-stage A partially cured thermosetting adhesive used in multilayer boards. During lamination, the application of heat and pressure completes the curing process.

ball grid array A circuit device package containing a large number of solder-coated ball terminations arranged in rows on the bottom of the package. When heat is applied during assembly, the solder on the balls is remelted, forming a connection to the circuit board.

bare board testing Continuity testing of an unassembled (unpopulated) circuit board.

base substrate The insulating material that forms the support for conductor patterns and components.

bed of nails test fixture A fixture consisting of a frame and a holder containing a field of spring-loaded pins that make electrical contact with conductors on the surface(s) of a circuit. Used for bare board and in-circuit testing.

bilateral dimensioning Linear dimensioning (as opposed to geometric dimensioning) that defines tolerances allowing variations from the specified dimension in both directions $(+/-)$.

bill of material A document that lists all electrical, mechanical, and supporting materials including their reference designations, quantities, associated part/find numbers, that are required to manufacture a PCA.

blind via A via that extends from an outer surface of a multilayer circuit board to at least one of the inner layers, but does not go completely through the board.

board construction   Defines the types and dimensions of materials, the layering sequence of the cross-sectional structure of a circuit board, and its finished thickness.

board detail drawing   A drawing that provides and describes all the requirements for fabricating a bare circuit board.

board extractor   A device that is used as a means of extracting a PCA from its mating connector without damage to its electrical components. It can be permanently mounted on the circuit board or provided as an external tool.

board profiling   A machining process for defining outline of a circuit board. See **board routing process.**

board routing process   A machining process for forming the outline of a circuit board. Pin routing uses a pin-guided template for manually profiling boards. NC routing utilizes programmable equipment to define a board profile. Both methods use cutters similar to end mills.

boundary scan   A diagnostic method that uses circuitry integrated in a functional electronic component to monitor the performance of that component and its surrounding interfaces.

bow and twist   Deviations from flatness requirements of a circuit board. A bow is measured from the top of a smooth arc to the same surface of the board if it were flat. Twist is a helical divergence from flatness.

breadboarding   "Quick and dirty" assembly and test of a circuit to validate its performance before committing it to implementation as a PCA.

bulk resistance   The resistivity (in ohm-cm or ohm-inch) through the cross-section of a conductive material.

buried via   A via that makes a circuit connection between internal layers of a multilayer board, and does not extend to either external surface of the board.

bus bar   A mechanical means of providing power and ground interconnections using conductive metal (usually copper) bars rather than etched conductors.

bus structure   A routed conductor pattern, usually used for point-to-point interconnection of power and ground.

bypass capacitor   Minimizes the effects of current variations in a power circuit caused by switching transients generated during circuit operation.

CAD (computer-aided design) A system used for automated design layout of PCAs.

CAE (computer-aided engineering) A system used as an automated tool for circuit design and schematic generation.

CAGE (commercial and government entity) code Identifies an individual company or corporation that manufactures a product. The CAGE code is usually marked on the PCA.

CAM (computer-aided manufacturing) system A system used to generate data for fabricating circuit boards and manufacturing finished assemblies.

centroid A point whose coordinates are the average of an associated part's dimensions (central point).

chassis The supporting frame or structure that houses PCAs.

chemical stability The ability of the characteristics of a material to remain unchanged by aging or variations in the environment.

chip-on-board (COB) An assembly process that places unpackaged components on a circuit board and interconnects them to the board by wire bonding or similar attachment techniques.

chip package A carrier in which an IC chip is mounted. The package interconnects the chip to the outside world and is sealed to provide environmental protection for the chip.

chip scale package (CSP) A package which requires a surface area for mounting on a circuit board that is no greater than 20 percent of the area of the die.

circuit The interconnection of a number of electrical elements and/or devices that together perform an electrical function.

circuit coupling The creation of a false signal in a circuit by a signal in another circuit. This is usually caused by radiated energy between adjacent conductors on the same or different layers.

circuit density The proportion of circuit elements and interconnections required for performing an electrical function to the allotted area of a circuit board.

circuit engineering The technical organization that is usually responsible for developing the PCA circuit design.

circuit fault Incorrect performance of a circuit resulting in a specific error. In digital circuitry, this could manifest itself as a logic error.

During functional testing, artificial faults are sometimes injected into a circuit to verify the tester's ability to correctly identify them.

circuit filter Protective circuitry designed to prevent transmission of unwanted current or voltage deviations during operation.

circuit frequency Usually pertains to the operating speed of a circuit and is a function of the types of components used, the dielectric properties of the circuit board, and the physical characteristics of the circuit conductors (dimensions/shape).

circuit net Defines individual component nodes that are connected together to define an isolated circuit.

circuit symbol Used in a schematic diagram as a graphic representation of a specific type of electronic device.

clinching The process of forming or bending a component lead following its insertion through a hole in a circuit board. The main purpose is to mechanically secure the part during the soldering process.

clock speed The switching frequency of the clock circuit in digital logic. It is principally determined by the rise/fall time required for the digital devices used in a circuit to change logic state (from 0 to 1 or vice versa).

cohesive characteristics See **adhesive characteristics.**

cold joint A solder joint to which insufficient heat has been applied, resulting in a dull granular structure of low mechanical strength.

component lead extension The distance a through-hole component lead extends beyond the surface of a circuit board after soldering.

conduction The ability of electrons to flow through a conductor. It is the reciprocal of resistance. Conduction is also a heat transfer mechanism in solid materials, involving transfer of kinetic energy within its molecular structure.

conductor A current-carrying interconnection path.

configuration control A method for ensuring that a specific version of data and drawings defines the correct requirements and physical description of the intended version of a PCA, and that the actual hardware conforms in all respects to the data and drawings.

conformal coating A thin insulating protective material that follows the contours of the objects coated and provides a barrier against deleterious effects of environmental conditions.

connector   A device that provides a mechanically pluggable interface for electrical terminations.

contact resistance   The resistance in the conductive path between two touching surfaces. It relates primarily to resistance across mating connector contacts.

convection   The mechanism for transfer of heat from a solid surface (such as a component) to a surrounding fluid (usually air). Natural convection is heat transfer to "still" air. Forced convection involves heat transfer to air that is moved by artificial means such as a fan.

copper-clad dielectric material   The basic material used for fabricating a printed circuit board, consisting of a flat reinforced dielectric to which is bonded copper foil on one or both surfaces.

copper foil   The conductive material used to form the interconnection pattern on a printed circuit board.

copper weight   A measurement of the thickness of copper foil, in terms of its weight in ounces per square foot of surface area (1-ounce copper is nominally 0.0014 in thick, $^1/_2$-ounce copper is 0.0007 in thick).

core   A supporting plane that is internal to a packaging and interconnecting structure. Multilayer circuit boards may contain unclad laminate cores, or special-purpose cores made of metal (aluminum, copper) to enhance heat dissipation or for other purposes.

cost-benefit trade study   The analysis of the cost of implementing each of an alternative (design/manufacturing/assembly/test) approach (or solution to a problem) versus the benefits of doing so.

critical signal paths   Conductors carrying signals that may be particularly sensitive to distortion by external signals, and require routing in specific locations on a board, or layout in a specific physical geometrical configuration.

crosstalk   See **circuit coupling.**

current rating   The maximum allowable continuous current that can be passed through a component or a conductor without causing degradation of performance.

cuts   Modification of a circuit board by separation of conductors on an external layer to break a circuit connection. Cuts should be a minimum width of 0.030 in.

dash number   Method for using part numbers to identify and control design modifications and interchangeability of assemblies. Utilizing a root part number with different dash numbers indicates that variations of the same functional design exist.

data file   A collection of information organized in a specific manner for a specific application.

data management function   The organization that is usually responsible for custodianship and control of technical data and documentation.

datum   The theoretical exact point, axis, or plane from which the location of geometric characteristics or features of a part are dimensionally established.

datum intersection (origin)   The point of intersection of the X and Y datums on a circuit board and the origin (0,0 point) of the layout grid.

de-coupling capacitor   See **bypass capacitor.**

dedicated service product   Equipment or system that must perform reliably over long periods of time and experience minimum downtime, such as communication equipment, computers, and online instrumentation.

delamination   Separation between layers of a multilayer board or between a conductor and laminate.

derating   Use of materials or components in a design at less than their rated characteristics (such as power dissipation or current-carrying capacity) to enhance the long-term reliability of the end product. Part manufacturers usually specify a derating factor to be used when a part is to be operated above a certain temperature.

design cycle   The entire technical activity associated with the design fabrication, assembly, test, and integration of a PCA.

design qualification   Verification through test and analysis that a PCA design will perform its required operational functions.

design reviews   Checkpoints established at critical points in the design process to verify the validity of the design and its associated data and documentation, and evaluate the producibility, testability, and projected reliability of the product.

design rules   Predetermined layout guidelines that are established to help ensure design optimization with respect to specified circuit performance parameters.

design standards   Layout processes, guidelines, and procedures that are widely used throughout the printed circuit industry.

dewetted surfaces   Metal surfaces (usually conductors) that have been covered with molten solder which has then receded, leaving a thin (less than acceptable), irregular coating of solder.

diametral   Pertaining to the diameter of a circle or hole.

dielectric constant   The ratio of the capacitance value of an insulating material to the capacitance value of the same geometry (area and thickness) using air as an insulator. It is a measure of an insulator's ability to store electrostatic energy.

dielectric material   An insulating material.

differential pair   Conductors carrying sensitive signals that should generally be routed in parallel with matched overall lengths.

digital circuitry   Circuitry that performs specific activities by implementing Boolean algebra functions (AND, NAND, OR, NOR, NOT).

digital clock lines   Conductors that carry a continuous stream of uniform pulses (0s and 1s) that establish the timing of operation of associated digital circuitry.

digital logic   See **digital circuitry.**

digital signal processor   An integrated circuit that electronically processes signals such as sound, radio, and microwaves by converting them from analog to digital signals.

digitizing   A method of capturing the *X-Y* coordinates of feature locations on a PCA layout and converting that data to a digital format.

dimensional origin   See **datum intersection (origin).**

dimensional tolerance   The total amount that a specific dimension is permitted to vary. The tolerance defines the maximum and minimum limits of the dimension.

dissipation factor   A measure of the absorption of electromagnetic energy passing through a dielectric material.

disturbed solder joint   A joint where physical movement has occurred during the solidification of the molten solder, resulting in a rough, granular, uneven solder surface and a joint of lower than average mechanical strength.

documentation/data release   The activity that takes place following final review and approval (signoff), when all drawings and design data are placed into a configuration/records control system.

dolls To-scale cutouts that represent physical parts to be mounted on a circuit board. They are used to perform component placement during a manual layout effort.

double sided A circuit board with conductive patterns on both external surfaces.

draw or flash A designation assigned to a photoplotting aperture. A flash aperture is the size and shape of the feature it defines on photosensitive film. A draw aperture creates the shape on film via software move commands transmitted to a photoplotter.

drawing Documentation that provides the configuration and requirements information needed to build a product.

drill data Information that specifies X-Y locations for all drilled holes, their sizes, and their plating requirements.

drill spindle runout The undesirable deviation from the theoretical center of rotation of a drill spindle due to its inherent mechanical tolerances. For high-density circuit boards, drilling total spindle runout should not exceed 0.0002 in.

drillout A method used to modify a fabricated circuit board or assembly by drilling through a conductor (usually internal) or plated hole to break the connection.

driver A signal source that generates an output strong enough to change logic levels of all devices (loads) attached to its net.

dry film material A photosensitive resist or solder mask material available as a film (as opposed to a liquid) that is applied to a circuit board during fabrication, using heat and pressure.

dummy traces Added nonfunctional conductors that help achieve plating balance. See **plating thieves.**

edge-card connector A pluggable connector specifically used for making nonpermanent interconnections to contacts plated on the edge of a circuit board.

ejector See **board extractor.**

electrical noise Variations from a nominal operational voltage or current value on a power or signal line. Circuits are usually designed to function within a noise tolerance band or allowable noise budget.

electrochemical migration Gradual movement of corrosive or conductive materials that may create unwanted circuit paths and cause performance degradation or failure. Migration is usually caused by expo-sure of existing contaminant residues or bare metal surfaces to a combi-nation of high humidity, elevated temperatures, and an electrical potential.

electroless plating   A process for chemically depositing metallic mate-rial on a nonconductive surface. It is primarily used for metallizing holes in a circuit board to prepare them for addition of electroplated metal.

electromagnetic radiation   Generation of a magnetic energy field that can cause interference with nearby circuitry.

electronic components   Passive and active devices that perform spe-cific functions in a circuit.

Electronics Industry Association (EIA)   A trade association repre-senting the U.S. high-technology community. It sponsors trade shows and conferences on behalf of its members and has been responsible for publishing numerous design standards.

electroplating   Direct deposition of metallic materials on a surface using an electrolytic process. It involves passing a current between an anode and a cathode through a conductive solution containing metal ions. The conductive surface of the laminate (circuit board) to be plated is the negative electrode (cathode).

electrostatic discharge (ESD)   The rapid transfer of a voltage poten-tial into a circuit or component. Depending on its sensitivity to ESD, the overstress can permanently damage a component.

EMI/EMC   Electromagnetic interference (EMI) is the detrimental effect stray, radiated electromagnetic energy has on the performance of a cir-cuit. Electromagnetic compatibility (EMC) involves the control and reduction of this energy.

emulator   A representation (hardware or software) of an electronic device or function that simulates its behavior. It is used to verify per-formance of a circuit design by analysis.

emulsion   A stable combination of two or more immiscible (unmix-able) materials suspended in a surrounding medium.

engineering change notice (ECN)   A document that describes and controls engineering design or documentation changes.

environment   The ambient conditions in which a PCA exists and functions, including temperature, humidity, altitude, vibration, shock, etc.

EPROM A programmable read-only memory device that allows data stored in it to be erased and new data input to it, usually by exposure to ultraviolet radiation.

etchback Chemical removal of dielectric material in the barrels of holes in a multilayer board. The purpose is to increase exposure of internal conductor areas (usually pads) to enhance physical and electrical contact with metal to be plated in those holes.

etching Chemical removal of material, usually associated with defining conductor patterns on a circuit board.

eutectic soldering process Application of heat to a tin-lead alloy (solder) until it reaches its melting temperature, at which point it will wet all adjacent surfaces. When allowed to cool, the solder will return to its solid state, forming an intermetallic bond (solder joint) with those surfaces.

Excellon data format A unique data format structure that is used to convert physical hole data (location, diameter, etc.) into a format that a software program can use to drive a numerically controlled (NC) drilling machine. It is considered a de facto industry standard.

false triggering An incorrect change of state of a digital device due to a spurious signal received by that device.

fatigue failure Mechanical failure of a material caused by application of repeated cycles of stress (force) and strain (movement) over a period of time. These forces may be due to vibration or caused by changes in temperature, and in PCAs may result in cracked plated through-holes, open solder joints, or board delamination.

fault isolation A test procedure for locating the area of a circuit that is causing a performance anomaly or failure.

Federal Communications Commission (FCC) An independent United States government agency, directly responsible to Congress, that is charged with regulating interstate and international communications by radio, television, wire, satellite, and cable.

Federal Stock Certificate Number (FSCN) The FSCN (code identification) number is a five-digit numerical code assigned to all manufacturers that have produced or are producing items used by the federal government.

FET A field-effect transistor is a unipolar device, which functions as a voltage amplifier.

fiducial   A feature that is etched on the outside surface of a circuit board or a panel of circuits at the same time as the conductive pattern is formed. It provides an optically measurable reference point for subsequent manufacturing and assembly processes.

field-programmable gate array (FPGA)   A device containing a large number of logic gates that can be interconnected internally by the user of the device to form an application-specific circuit.

fill area   A large conductive area such as a ground or power plane.

find number   An item number that cross-references a part callout on an assembly drawing to its entry in a parts list or bill of materials.

fine-pitch   Refers to components with terminations on less than .025-in centers.

finite-element modeling (FEM)   A method of using a software program to simulate the response of a PCA to various mechanical or thermal conditions. A mathematical model of an assembly is constructed, exposed to mechanical or thermal stimulation, and analyzed for its response to those inputs.

flex circuit   Printed circuitry that utilizes flexible rather than rigid laminate material.

flexible buffer material   If devices packaged in a brittle material (such as glass diodes or ceramic ICs) are coated with rigid conformal coating material, temperature cycling may cause the packages to crack due to the difference in the thermal coefficients of expansion of the two materials. This may be prevented by applying a resilient material (such as silicone rubber) to the part as a buffer between it and the conformal coating.

flip-chip device   A leadless, monolithic, circuit element (chip) that is electrically and mechanically interconnected to a conductor pattern on a board through the use of conductive bumps on the chip. The bumps are formed on the active surface of the chip, which is turned over (flipped) for attachment.

flux   A material used in conjunction with soldering that removes oxidation on surfaces to be soldered and prevents reoxidation during the formation of a solder joint.

flux residue   A corrosive, conductive byproduct of the soldering operation that must be completely removed to prevent future degradation of circuit performance, or failure.

footprint   The hole, pad, and conductor pattern associated with a specific electronic component package configuration.

form, fit, or function   Interchangeability classifications that determine the necessity for re-identification (renumbering) of a product when design changes are implemented. If the physical form of a PCA, its ability to fit in the same place as the previous design, or its functional operation changes, it should be considered a new product.

functional testing   Verification of the proper performance of a PCA by connecting it to a tester that can simulate the interface between the PCA and its final application.

geometric dimensioning   A method that defines the location and associated tolerances of a mechanical feature in terms of its allowable position with relation to its theoretical (true) position.

Gerber data   A unique data format that converts physical feature information (location, shape, etc.) into a format a software program can use to drive a numerically controlled photoplotting machine that produces printed circuit artwork.

glass transition temperature (T_g)   The temperature at which the mechanical and electrical properties of a plastic material (laminate dielectric) begin to degrade.

gold plate removal   When solder is applied to a gold surface, a brittle intermetallic substance is formed at the interface, which may eventually cause a failed solder joint. Gold plating is usually stripped from a conductive surface prior to soldering to prevent this.

grid origin   The 0, 0 point of a layout grid, usually located in the lower left corner of a board. See **datum intersection (origin).**

gridless (shape-based) router   The type of CAD autorouting program that has the flexibility to find paths for conductors based on the shape of surrounding geometric features (such as conductor widths and spacings), instead of being restricted to a predefined grid.

ground and power planes   Continuous metal surfaces that usually cover all or a large portion of a board and distribute power to circuitry on the board. A ground plane may also be used as an electromagnetic shield and as a reference plane for high-frequency (stripline) circuitry.

ground loops   Undesired current flow due to differences in potential across a ground plane.

hard-wired interconnections Circuit connections using wire as opposed to etched interconnections.

heat removal mechanisms Heat may be transferred from a surface by any or all of three mechanisms: conduction, convection, and radiation. See **conduction, convection,** and **radiation.**

heat sink A mechanical device that aids in removal of heat from electronic components.

high-density interconnections Relates to the average amount of circuitry packaged in a given area of a PCA, in terms of linear inches of conductor and the number of component I/Os terminated in that area. Boards with average I/O counts in the range of 200 to 300 per square inch of area or greater, should be treated as high-density designs.

high-speed circuitry Circuits operating at speeds of 50 MHz or greater are normally considered to be high speed, although special consideration to conductor placement and geometry starts at circuit speeds above 25 MHz.

hole breakout A condition in which a hole is not completely surrounded by its associated land.

hybrid circuit A functional circuit assembly containing a substrate (usually ceramic) on which are deposited various combinations of thick or thin film conductors and passive components. Semiconductor and passive chip components are physically attached to the substrate and interconnected to the rest of the circuitry.

I/O connectors Connectors that are the interface between a PCA's input/output signals and the outside world.

impedance The inductance L of a circuit interconnection system divided by its capacitance C, stated as Z_0 in ohms.

in-circuit testing Test of an assembled PCA to verify proper operation of the electronic components attached to the circuit board. The terminals of each part are accessed by a bed of nails fixture, which effectively isolates it from the rest of the circuitry and allows it to be tested as a stand-alone device.

industry standard Processes, procedures, guidelines, and data formats that are widely used and recognized throughout the printed circuit industry.

infant mortality A marginal part, circuit board, or solder joint that was not caught by production test or inspection, and fails a short time after being put into operation.

ink pattern Applied to the surface of a circuit board through a screen or stencil, ink can be used to add markings or as a resist to define an interconnection pattern.

IPC—Association Connecting Electronics Industries A leading printed wiring industry association that develops and distributes standards, as well as other information of value to printed wiring designers, users, suppliers, and fabricators.

Institute of Electrical and Electronics Engineers (IEEE) An organization that promotes engineering processes by creating, developing, integrating, sharing, and applying knowledge about electrical, electronic, and information technologies and sciences for the benefit of the profession.

insulation resistance The ratio of the potential applied between two points of an insulator to the current flow between them.

insulator A nonmetallic material designed to prevent current flow.

integrated circuit (IC) An electronic component containing a number of interconnected circuit elements that enable the component to perform a specific function.

integration The marriage of an operational PCA with the system or product with which it is to function.

intelligent data Electronic data that contains physical or functional information about an item (such as an electronic part), and also links to other information.

interchangeability The characteristic of a design that allows direct replacement of one item with another without requiring any modifications.

interconnection escapes Conductor paths provided in a device footprint for access to the part's terminals.

International Electrotechnical Commission (IEC) The IEC is a world organization that prepares and publishes standards for electrical, electronic, and related technologies.

International Organization for Standardization (ISO) The stated mission of ISO is to promote the development of quality standards and related activities in the world to facilitate the international exchange of goods and services.

ir drop   The voltage drop through a conductor.

ir emitter   A source of light energy in the infrared spectrum.

ir reflow   Use of infrared energy to bring solder to its melting point.

Joint Electronic Device Engineering Council (JEDEC)   Develops and publishes configuration standards for semiconductor device packages.

jumper   A wire that forms a discrete electrical connection between conductive areas on a circuit board's external surface.

junction temperature   The operating temperature of the active element of a circuit component.

keep-out area   A predetermined region on a circuit board where specific items (conductors, parts, holes, etc.) cannot be placed during layout.

keying   A mechanical method of preventing incorrect interconnection of mating components (such as connectors).

laminate   See **copper-clad dielectric material.**

lamination   The process of fabricating a circuit board by using heat and pressure to glue together a number of interconnection layers to form a single multilayer assembly.

land   A conductive area in a circuit pattern used for the connection and/or attachment of component terminations (usually by soldering, or as a contact point for a test probe).

land pattern   A combination of pads that is used for the mounting, interconnecting, and testing of a specific type of component.

layout   Depicts the physical size and location of electronic and mechanical components on a circuit board, and the routing of conductors that electrically interconnect the components. Information is provided in sufficient detail to allow the preparation of documentation and artwork for fabrication, assembly and test of a PCA.

layout grid   A lattice of orthogonal lines spaced in standard increments (typically 25 or 50 mils). Components, plated and nonplated holes, surface mount land patterns, and other features are usually located at the intersection of these grid lines during the layout of a circuit board.

layout rules   Rules established, based on the design type and perform-ance requirements, that determine component placement, conductor routing, layer stack-up, etc.

lead forming   The process of bending component leads so that they may be inserted into holes or surface-mounted on a circuit board.

library   A structured catalog of related items (such as schematic sym-bols or component part descriptions) that contain all the information about the items that is needed for their use in a design.

lifted pad   A pad or land that has partially separated from its base material.

loads   Digital devices attached to a net that will have their logic levels changed by a driver on that net.

logic families   A group of device types that share the same basic oper-ating characteristics and parameters when processing digital signals (TTL, ECL, etc.).

lot and date code   Items manufactured in a group using the same materials and processes are given the same, unique lot code. Items manufactured or completed on the same date are given the same date code.

maintenance and field support   The technical organization usually responsible for maintaining and repairing an item that is in the hands of the end user.

manual pin routing   See **board routing process.**

manufacturability   A measure of the producibility of a PCA design.

manufacturer's data specification sheet   It provides electrical and mechanical data about a part and is used as a source of information for library inputs, design activities, and material procurement.

manufacturing data   Information consisting of photoplotting files, drill files, pick-and-place files, bare and loaded board testing files, sup-port drawings, and bills of materials required to do the actual fabrica-tion, assembly, and test of a PCA.

manufacturing engineering   The technical organization that is usually responsible for planning and implementing PCA production activities.

markings   Information imprinted on parts and circuit boards, such as reference designations, part or serial numbers, revision level, orientation or polarization symbols, bar codes, electrostatic discharge (ESD) sensitivity, etc.

mass soldering   The process of forming all solder joints on a PCA simultaneously (wave soldering, dip soldering, oven reflow, vapor phase soldering).

master artwork set   An accurately scaled, 1:1 pattern that is used as the source for producing working artwork films for circuit board fabrication.

master pattern drawing   A document that shows the dimensional limits or grid locations applicable to any or all parts of the circuit board. This also includes the arrangements of conductive and nonconductive patterns or elements; the size, type, and location of holes; and other information necessary to describe the product for fabrication.

material panel   Laminate, prepreg, and copper foil materials used for manufacturing circuit boards that are produced in standard sheet or panel sizes.

maximum/least material concept   Used in geometric dimensioning and tolerancing, maximum material describes the condition of a mechanical feature within a stated limit of size (minimum hole size, maximum shaft diameter). Least material describes the minimum size of a feature (maximum hole size, minimum shaft diameter).

mean time before failure   A mathematical function that describes the probability of failure of a product after a specified amount of operational time.

measling   An internal defect in laminate material involving separation of reinforcing material fibers from surrounding resin, usually occurring where the fibers cross.

mechanical engineering   A technical organization that is usually responsible for defining the physical size and shape of a circuit board, input/output termination methods and locations, area and volume restrictions, accessibility requirements, types of mounting/removal hardware, and environmental management methods.

mechanical shock and vibration   External forces imposed on assembly by cyclical energy inputs (vibration) or a single, sudden high-energy input (shock).

media   The type of material used for product data and documentation, such as electronic (tape, CD, disk) and drawings (paper, Mylar, film).

metal foil   See **copper foil.**

microprocessor   An integrated circuit containing all the functions required to operate as a computer.

microstripline   A type of high-frequency transmission line configuration that has a specific, characteristic impedance value. It consists of a conductor placed in a precise relationship with a ground or reference plane and surrounded by dielectric materials.

microwave   Circuitry that operates at frequencies between 1 GHz and 300 GHz.

microwave integrated circuit (MIC) module   A circuit assembly designed to function in the microwave frequency range. It usually contains a substrate with stripline circuitry and active chip devices, all sealed in a metal case for shielding purposes.

mixed analog/digital   A PCA containing both analog and digital circuitry.

mixed technology   When both through-hole and surface-mounting component types are used on the same PCA.

modeling   A design analysis method using a software description of an item (electrical or mechanical) to simulate its operation in response to a set of stimuli.

motherboard   A circuit board assembly used for interconnecting a group of plug-in electronic assemblies.

multichip module (MCM)   A microcircuit package containing bare semiconductor, chip-scale devices (ICs) that are mounted and interconnected on a mini-PC board or substrate.

multilayer board   A circuit board containing internal conductor layers

multilayer lamination   See **lamination.**

NC equipment   Any machine whose activity is commanded by instructions that are input to a programmable controller.

NC fabrication equipment   Numerically controlled machine tools such as routers and drilling machines.

neckdown   The localized reduction of a conductor's width to allow it to be routed through tightly spaced patterns.

net   An independent set of circuit nodes on a schematic that are connected together. Each net is given a unique (alphanumeric) name to differentiate it from the other nets in the schematic.

net list   An alphanumeric listing of all nets in a PCA schematic.

net names   See **net.**

node   An individual component termination (pin), test point, or I/O within a circuit net.

nonfunctional pad   A land on an internal or external layer that is not connected to an active conductive pattern on that layer.

nonrecurring cost   The one-time (hopefully!) cost of design and development activities prior to starting production of a PCA.

off-grid   Circuit board features that are not located on a grid intersection.

opacity   The level of transmissibility of light through a material. It refers to the ability to read markings covered by conformal coating or solder mask material on a PCA.

operating voltage   The nominal voltage a circuit component requires to function properly.

operational amplifier   A circuit device that increases the power level of an analog signal.

operational environment   See **environment.**

optical alignment mark   See **fiducial.**

orientation marking   Information imprinted on a circuit board that provides component location information for installation of polarized or multileaded parts.

original equipment manufacturer (OEM)   Manufacturer of a product that is intended to perform a function when operated by an end user, as opposed to a manufacturer that builds components that go into such equipment.

outgassing   The breakdown of material resulting in diffusion of gaseous contaminants to its surface that are then transmitted into the surrounding atmosphere. It can be caused by exposure to high operating temperatures and/or altitude (low external pressure).

overstressed   Material or component exposed to operating conditions beyond specified limits.

oxidizing   A chemical process involving combination of a material (usually a metal) with oxygen to produce a substance with substantially different physical and electrical characteristics from the base material.

package I/O   An individual component's circuit terminations.

package material The type of material used for a specific type of electronic component (plastic, ceramic, metal).

PAD See **land.**

panel A rectangular sheet of base material or metal-clad material of predetermined size that is used for the processing of one or more circuit boards.

panel plating The process of electroplating copper on all conductive surfaces and holes of a circuit board. A circuit is defined by applying a resist pattern to the plated surfaces and etching away all copper not covered by resist.

panelization The placement of multiple circuit board patterns on a single panel.

panelized circuit pattern Defining multiple circuit boards on a single laminate panel so that they may be processed simultaneously.

part footprint A standard conductor, land, and hole pattern that is unique to a specific type of component. It is used for mounting, soldering, and interconnecting those components on a circuit board.

part library A structured catalog of individual-component-part physical descriptions that contains all the information about each part needed during the layout of a circuit board.

partitioning The allocation of functional circuitry to a PCA based on performance requirements and the physical space ("real estate") available for part placement and interconnection.

parts Individual items used in a design such as electronic components, mechanical hardware, etc.

parts list A tabulation of all parts and materials used in the construction of a PCA. See **bill of material.**

passive component A part that exhibits a fixed or controlled value and performs an elementary function in a circuit, such as a resistor, capacitor, inductor, or conductor.

pattern plating The process of electroplating metal only where a conductor is to be formed. A circuit is defined by applying a negative resist pattern to conductive surfaces and plating etch-resistant material (usually tin-lead) in the openings in the resist. The resist is then removed and the unwanted copper etched away leaving the conductor pattern.

PCA Printed circuit assembly.

peel strength   The strength of the bond between copper foil and the base laminate to which it is attached.

photo imageable resist   A photosensitive material available as a film or a liquid that is applied to the surface of a circuit board during fabrication using heat and pressure. Collimated light passed through an artwork film defines an image in the material. Development of the material leaves an etch/plating-resistant pattern that is used to define the conductors on that surface of the circuit board. Photoresists are either positive (areas exposed to light remain when the image is developed) or negative (areas exposed to light are removed when the image is developed).

photoplot data   Electronic data generated for use by photoplotting equipment.

photoplotter   NC-controlled equipment used for generating an artwork image on a light-sensitive emulsion coated on a stable material (plastic film or glass plate).

photoplotting   A process that creates an image on a photosensitive material by a controlled light beam.

phototools   See **artwork.**

pick-and-place   The automated assembly process that uses NC equipment to precisely place electronic parts on a circuit board prior to soldering.

pin grid array   A circuit device package containing a large number of pin terminations arranged in rows on the bottom of the package.

plated hole   A metallized hole in a printed circuit board that makes electrical connection between conductors on different circuit layers. It is formed by plating metal on the internal hole wall.

plating thieves   Nonfunctional metal areas on a surface to be electroplated. Their purpose is to balance the current density during plating to ensure uniform buildup of plated material. See **dummy traces.**

polarity marks   See **orientation marking.**

positive/negative image   A developed photographic film or plate on which the defined object or pattern is transparent when it is a negative image. If the pattern or object is opaque, it is a positive image.

postprocessing   Conversion of layout information into data files having formats that can be used by equipment employed to fabricate, assemble, and test a PCA design.

power density   The distribution (or concentration) of power dissipation of electronic components and interconnections across the surface area of a PCA.

precision artwork master   See **artwork.**

precision drilling   The process used to produce accurately located holes in a circuit board with closely held diametral tolerances.

prepreg   See **B-stage.**

pretinning   The process of applying a fresh coat of solder (tin-lead) to component leads prior to mounting them on a circuit board. This is done to enhance solderability by removing/replacing oxidized material on the leads.

primary side of PCA   The side of a circuit board on which the most complex or highest number of components are mounted.

printed circuit assembly   An assembly designed to perform a specific function, consisting of a printed circuit board to which is mounted and interconnected electronic components.

printed circuit board   A general term for a fabricated substrate containing a defined interconnection pattern on which is to be mounted electronic components and mechanical hardware.

printed circuit design   The process that depicts the printed wiring base material, the physical size and location of electronic components and mechanical parts, and the routing of conductors that electrically interconnect the components.

printed circuit fabrication   The process of manufacturing a bare circuit board.

printed wiring board (PWB)   See **printed circuit board.**

product life cycle   Encompasses fabrication, assembly, test, storage, transportation, and operation of a product.

programmable read-only memory (PROM)   See **EPROM.**

pull-up and pull-down resisters   Resistive components that are used as terminations on transmission lines to reduce or eliminate signal reflections due to line discontinuities.

purchasing function   The technical organization that is usually responsible for procuring components and materials.

quality assurance function   The technical organization that is usually responsible for verifying and validating the performance and reliability of a product and ensuring that it meets all of its specification requirements.

radial lead A component terminal that protrudes raylike from the body of a component.

radiation The mechanism for transfer of heat from a solid surface (such as a component) by electromagnetic transmission.

random access memory (RAM) A device that stores information, which can be both written and read many times. Any part of the memory can be accessed directly through an address. Data in RAM cells can be erased or changed by being overwritten or by removal of power from the device.

rat's nest A graphic display produced by a CAD system that shows all interconnections between circuit nodes on a layout as a set of straight lines. Ideal usage of this display is for optimizing part placement.

read-only memory (ROM) A device that is used for permanent storage of information. Data can be read many times, and remains when power is removed from the device.

recurring cost The cost that is incurred for each item produced, including material and labor.

reference designator An alphanumeric identifier assigned to each electronic component in a circuit. The alpha part defines the type of component (R = resistor, C = capacitor, etc.) and the numeric part is the sequential number assigned to a component (R-23 is the 23rd resister used in the circuit).

reflections The undesirable return of signal energy due to a discontinuity in a transmission line in which the signal is traveling.

reflow soldering The remelting and resolidification of solid or paste solder to form an electrical connection.

registered land pattern (RLP) A specific component pattern geometry defined by the IPC that has been used and accepted as an industry standard for that type of part.

registration The degree of positional alignment to each other of internal and external features (primarily pads and holes) of a circuit board.

reinforcement Material embedded in the resin of a laminate to provide additional mechanical strength. Typical materials are glass cloth, random glass fibers, paper, and a variety of high-strength plastic fibers.

reliability   The probability that an item will function under a specific set of conditions, for a stated period of time, without failure or unacceptable degradation of performance.

resin   A nonconductive plastic material, such as epoxy, polyester, or phenolic, used to produce printed circuit laminates.

resist   Any material used to define a pattern by preventing the products associated with a manufacturing process from attacking (etching) or adhering (plating) to the surface covered by the resist.

revision letter   Sequential alphabetic designators used on documentation and data to identify and control changes. The revision letter of a drawing should be advanced (A to B, etc.) any time the drawing is modified, to differentiate it from the previous version.

rf   Radio frequency circuitry operating in the range between 10 KHz and 1 MHz.

rigid-flex   A circuit board combining both rigid and flexible dielectric materials in a single assembly.

ringing   Short-term spikes in a signal. Usually related in digital circuitry to transients generated when a gate changes (switches) logic states.

risk   The probability of occurrence and potential negative impact of a decision or action on downstream activities.

root number   The portion of an assembly or part number that identifies its unique type or function (for PA123456-001, PA123456 is the root number)

routing   Establishment of paths on a board for circuit interconnections.

routing channel   The space available to route conductors between existing circuit features (pads, vias, holes, prerouted traces, etc.).

rule class   A set of predetermined layout rules [spacing, voltage, conductor size(s), current, isolation, etc.] that are associated with a specific type or class of circuitry. For example, there may be a different rule class used on a layout for analog circuitry than for digital circuitry on the same PCA.

schematic diagram   A drawing which shows, by means of graphic symbols, the electrical connections, components, and functions of a specific circuit arrangement.

schematic symbology   A graphic diagram utilized to represent a specific type of component and its terminations on a schematic diagram.

screening A process for transferring an image to a surface by forcing a viscous liquid material (such as ink, resist, or solder paste) through a screen with a squeegee. Also refers to the process of inspecting or testing a group of materials or parts to weed out noncompliant items.

semiconductor device An electronic component containing an active circuit material whose conductivity can be varied by a variety of external inputs (voltage, light energy, heat, etc.).

sequencing interconnections Controlling the order of interconnection of nodes in a net during routing to enhance the performance of sensitive circuits.

shadowing solder joints Blocking the proper formation of a connection during wave soldering. It is usually caused by an adjacent component.

shielding An electrically conductive physical barrier designed to reduce the detrimental interaction of electromagnetic fields upon devices or circuits. See **EMI/EMC.**

signal coupling See **circuit coupling.**

signal edge The forward or trailing part of a signal pulse that causes a gate to change logic states.

signal flow The movement of data through a circuit.

signal integrity The specified purity (lack of distortion) of a signal transmitted through a circuit required for proper operation. Design of a circuit, selection of electronic parts, their physical placement on a board, and the location and physical configuration of the interconnections all have a significant effect on signal integrity, especially for high-speed circuitry.

signal layer An interconnection layer on a circuit board devoted exclusively to the routing of signal traces.

signal propagation delay The time it takes a signal to travel (propagate) in a line, compared to its theoretically possible speed. It is mainly a function of the characteristic impedance of the line, the dielectric constant of the surrounding material, and the type and quantity of the components attached to the line.

signal rise and fall times The time it takes the edge of a signal pulse to reach a value that will cause a gate to change states.

signal timing The speed at which a signal that causes digital devices to switch (change states), travels in a circuit. See **clock speed.**

signal transmission line   A conductor that is configured to have a specific impedance value. See **microstrip line** and **stripline.**

simulation   Use of a computer program that duplicates the characteristics of an entity. Its purpose is to verify the performance of a design before committing it to a hardware implementation. See **modeling.**

single-sided   A circuit board with a conductive pattern on one external surface only.

Society of Automotive Engineers (SAE)   A technical society that prepares and publishes material and process specifications, standards, technical papers, and books.

software engineering   The technical organization that is usually responsible for developing and designing operational software.

solder   A tin-lead alloy that is melted and allowed to resolidify to form an electrically conductive joint between a component lead and a printed circuit conductor.

solder dam   A neckdown (narrowing) of a conductor that restricts the flow of molten solder. Its main purpose is to ensure that the proper amount of material remains at the solder joint that is being formed.

solder leveling or fusing   The process of remelting plated solder (tin-lead) on the surface of a circuit board to control its thickness, reduce granularity, and eliminate harmful oxidation.

solder mask   A coating material used to shield selected conductor areas from the application of solder.

solder mask over bare copper (SMOBC)   The application of solder mask material over unplated copper conductors. There is sometimes a concern that solder plating on conductors that are covered by mask material will remelt and flow during the board soldering operating, damaging the mask and the board itself. SMOBC eliminates that possibility.

solder paste   Fine particles of solder, with additives to promote wetting and to control viscosity, tackiness, slumping, drying rate, etc., that are suspended in a viscous flux material. Solder paste is screened on a surface of a circuit board, and remelted to form solder joints for surface-mounted components.

SPICE   A computer program that is used as a tool to analyze the performance of analog circuitry.

stackup   Defines the construction and layup of a circuit board.

static and dynamic thermal conditions   Heat flow conditions/ values that remain constant are considered thermally static. Heat flow conditions that are continuously changing are thermally dynamic.

stripline   A type of high-frequency transmission line configuration that has a specific, characteristic impedance value. It consists of a conductor placed in a precise relationship between two parallel ground planes and surrounded by a dielectric material.

stub   A short projection or branch from the main path of a conductor. In high-frequency operation a stub can act as a discontinuity in a circuit, causing degradation of performance.

surface mount   The electrical connection of components to the surface of a conductive pattern without using component lead holes.

synthesizer   A computer program that allows an engineer to specify the logic operations that a design is expected to perform. The synthesizer extracts the equivalent logic circuit functions from a library and connects them together as specified by the engineer to form a complete circuit.

system integration   See **integration**.

systems engineering   The technical organization that usually defines the performance requirements for a system or subsystem and ensures that the requirements are satisfied by the design that implements them.

tape-automated-bonding (TAB)   Attachment of chip devices to a series of flexible lead frames mounted on a strip of carrier film material (tape). When separated from the strip, each frame interconnects its chip to the next level of assembly.

temperature gradient   The rate of change in temperature.

terminator   A device (usually resistive) attached to the end(s) of a transmission line to prevent or reduce signal reflections that could affect the performance of the circuit.

test coupon   A pattern put on a panel or circuit board that undergoes the same fabrication processes as the board. It is used to verify quality and conformance and to perform product acceptance tests.

test engineering   The technical organization that is usually responsible for the testing of assemblies that are being manufactured.

test pad   Designated points of access to a circuit or component for testing purposes.

test pattern   During testing of a PCA, patterns of signal faults are inserted into the circuitry, and its response is observed to determine if the faults were recognized appropriately. See **circuit fault.**

theory of operation   A description of how a circuit design is supposed to function.

thermal coefficient of expansion (TCE)   The rate of expansion of a material as its temperature increases (measured in ppm/°C).

thermal grease   A viscous material (such as silicone grease) that is inserted between two surfaces to enhance heat transfer between them.

thermal relief pad   A land configuration used for connecting plated through-holes to large conductive areas (power or ground planes) which become heat sinks during a soldering operation. It prevents excessive loss of heat of solder in the hole, which could result in formation of an inadequate joint.

thermode   A heated bar that is used during rework/repair of a PCA to remove a component by simultaneously desoldering multiple component leads

thick film hybrid   See **hybrid circuit.**

thin film hybrid   See **hybrid circuit.**

through-hole   A hole that extends though the entire circuit board. It may or may not be plated, depending on its function.

through-hole via   A plated hole made to extend completely through a circuit board hole for the sole purpose of connecting conductors on one or more layers.

tombstoning   A component mounting and interconnection defect condition where a leadless device is standing on end after a soldering operation, and its metallized terminations on only one edge are soldered to lands on the circuit board.

tooling hole   A feature in the form of a hole in a circuit board or fabrication panel that is used as an aid for manufacturing.

traces   See **conductors.**

transmission line effects   See **signal transmission line.**

true position   The theoretically exact location of a feature or hole established by a basic dimension.

true positioning tolerancing   See **geometric dimensioning.**

two-part connector   A device that provides a mechanically pluggable interface for electrical terminations. One-half of a connector pair is mounted on a circuit board and the mating half is electrically connected to the rest of the system.

U.S. Department of Defense (DoD) and NASA   The Department of Defense and NASA are government agencies that impose very stringent specification requirements on PCA designs.

Underwriter's Laboratory (UL)   An independent, not-for-profit product safety testing and certification organization.

Unix   One of the first computer operating systems to be written in C, a high-level programming language.

vapor phase   A reflow soldering method. It is based on the exposure of an assembly to a material that releases heat sufficiently high to melt preapplied solder when it changes phase from vapor to liquid.

via   See **through-hole via, blind via,** and **buried via.**

via tenting   Covering a via with a masking material, such as a dry film polymer coating (solder mask), B-stage (prepreg), etc., in order to prevent hole access by process solutions, solder, or contamination.

voltage breakdown   The level of voltage potential across a normally nonconductive material where current begins to flow in the material or on its surface. This may be accompanied by a sudden electrostatic discharge, which can severely damage the material.

warpage   Also referred to as bow and twist, this is the deviation from the flatness requirement of a PCA.

wave soldering   A process for producing solder joints on an assembled circuit board by passing it over the surface of a continuously flowing and circulating wave of molten solder.

wedge lock   Mechanical hardware utilized for constraining the edges of a circuit board assembly in a mechanical chassis. It reduces the possibility of damage due to board vibration and improves the heat conduction path between the assembly and the chassis.

wiring harness   A prefabricated bundle of wires.

workmanship requirements   A set of quality standards that must be met by a manufactured PCA.

x-raying   An inspection process used mainly for determining the alignment of internal features (pads, conductors, etc.) of a multilayer board. It may also be used to determine the quality of solder joints that cannot be inspected by direct visual means.

yield   The percentage of acceptable products obtained from a total population of the products.

zener diode   An active component through which current flows more easily in one direction than the other. It is designed to work in a reverse biased condition, and is used primarily for voltage regulation.

APPENDIX B

B.1 CCA Failure Mechanisms and Models

Failure is defined as the inability of a system to perform its intended function. Failures in circuit card assemblies (CCAs) result from a complex set of interactions between the stresses that act on a CCA and the materials of a CCA. *Failure mechanisms* are the processes by which a material or system degrades and eventually fails. Three basic categories of failure include overstress (that is, stress-strength), wearout (that is, damage accumulation), and performance tolerance (that is, excessive propagation delays). Overstress and wearout failure categories are commonly related to irreversible material damage; however, some overstress failures can be related to reversible material damage (that is, elastic deformation). Overstress and wearout failure are characterized by a damage metric. In general, damage metrics for overstress failures are defined in terms of a pass or fail criteria. The damage metrics for wearout failures define how much useful life is consumed by an imposed environmental condition. Performance-tolerance failures may be associated with stress conditions that produce reversible changes in material properties. As a result, failures occurring at elevated stress levels may not occur during normal operation. Performance-tolerance failures are characterized by a performance metric. Failure occurs if the performance metric is outside of a specified tolerance range. Performance-tolerance failure mechanisms are usually associated with the failure of integrated circuits and not the CCAs (for example, excessive propagation delay in integrated circuits due at high temperature or excessive thermal transients due to inadequate diffusivity).

---

*Reprinted with permission from T. J. Stadterman and M. D. Osterman *in* C.A. Harper, Ed., *High Performance Circuit Boards,* McGraw-Hill, New York, 2000, Chap. 9, pp. 9.36—9.89

349

In the physics of failure (PoF) approach, failure is related to the physical structure of the system under consideration and the environmental/operational loads. The failure modeling generally involves an interrelated two-step process. The first step is the evaluation of a stress metric. The stress metric, defined by the damage or performance metric, is generally a function of the stress condition, system geometry, and material properties (that is, the stress metric characterizes a physical situation). The second step is determining the damage metric, which is related to the material system or interface under consideration. The damage metric, using the stress metric as an input, determines the amount of useful life consumed by a physical situation. It is highly desirable to separate the stress metric and the damage metric, which allows easier solution to the problem. If the stress and damage metrics are closely coupled, their relationship can be confusing. In some cases, the stress metric is an artificial parameter that may be mistakenly treated as an actual stress value.

This section will focus on overstress and wearout failures that are more predominant in CCAs. These failure mechanisms can be categorized by the nature of the stresses that drive or trigger them. These categories include the following:

- *Mechanical.* Failures that result from elastic and plastic deformation, buckling, brittle and ductile fracture, fatigue-crack initiation and propagation, creep and creep rupture.

- *Thermal.* Failures that result from exceeding the critical temperatures of a component such as glass-transition temperature, melting point, or flash point.

- *Electrical.* Failures that include those due to electrostatic discharge, dielectric breakdown, junction breakdown, surface breakdown, surface and bulk trapping, hot electron injection.

- *Radiation.* Failures that may be caused by radioactive contaminants and secondary cosmic rays.

- *Chemical.* Failures that result from chemical environments which act as catalysts to corrosion, oxidation, or ionic surface dendritic growth.

This breakdown of failure mechanisms allows an analyst to consider stress metrics that must be evaluated to determine the potential for failure. To completely define the stress metric, the environmental/operational loads must be related to the geometry and materials under investigation. This can be achieved by categorizing failures in CCAs by common structures or building blocks. Common CCA structures include the following:

- Components (active and passive)
- Permanent interconnects
- Metallization
- Contacts

The failure information related to each structure is included in this section. Component failure is particular to the technology and physics of the specific devices and is beyond the scope of this chapter. This section will focus on failure mechanisms and models of permanent interconnects, PWB metallization, and contacts. Before addressing these specific failure mechanisms and model, the next section will describe generic damage models.

B.1.1 Damage models and metrics

Damage metrics approximate the probability of a failure occurring and are based on the physical conditions that produce failure. Damage is based on the assumption that failure results from irreversible degradation of the material which accumulates over time. The underlying assumption is that damage is additive, and this assumption is used to estimate the damage to permanent interconnects as a result of repeated stress reversals. The common damage metric for failures resulting from cyclic conditions is expressed as the ratio of applied (exposed) cycles to the estimated number of survivable cycles at the current exposure level. This is mathematically expressed as

$$\mathrm{DM} = \frac{N_{\mathrm{applied}}}{N_{\mathrm{life}}} \tag{B.1}$$

where N_{applied} is the exposure time of the failure-inducing condition and N_{life} is the estimated time to failure for the failure-inducing condition. If $N_{\mathrm{applied}} = N_{\mathrm{life}}$, the probability of failure is assumed to be one. Using this approach, failure can be evaluated for multiple environmental conditions. In this case, the damage metric is defined as

$$R = \sum_i \frac{N_{applied_i}}{N_{life_i}} \tag{B.2}$$

This ratio is referred to as Miner's ratio. Cycles to failure is commonly used for fatigue-related failures. For failure resulting from a sustained condition, such as creep rupture, the damage metric is generally defined as time to failure. The damage metric in this case is the application time over the estimated time to failure.

One of the inputs to the failure model is the stress metric. The term stress metric is introduced to distinguish it from the physical stress. In many cases, failure models blur the lines between actual physically measurable parameters and metrics that are used in modeling physical failures. The failure model is generally closely coupled with the method of characterizing the stress history.

B.1.2   Permanent interconnects

One of the greatest concerns related to CCA reliability is failure of package-to-board interconnects. For most modern packages, package-to-board interconnects provide a structural and an electrical interface between the component and the PWB. Solder is the most common method of interconnection. Failure mechanisms related to permanent interconnects include fatigue, creep rupture, mechanical overstress, intermetallic growth, and corrosion. Fatigue is generally considered to be the primary failure mechanism and can occur due to temperature cycling and/or mechanical vibration. While important, elevated temperature, chemical contamination, and moisture generally play a secondary role in precipitating interconnect failures. The fatigue failure of permanent interconnects will be addressed in the following sections.

B.1.2.1   GEOMETRY AND MATERIAL CHARACTERIZATION.   Fundamental to permanent interconnect failure modeling is the understanding of the structures and the material used. In order to streamline production of the electronic system, electronic components are packaged in standard formats. Modern components have a multitude of package types or families. The standardization of package format is conducted by the Joint Electron Device Engineering Council (JEDEC) in conjunction with the Electronic Industry Alliance (EIA). JEDEC currently recognizes over 200 unique package styles, and new package formats are still being introduced. While there may be many similarities between package style, there are also important distinctions. The clearest delineation of package styles is between insertion and surface-mount technologies. Insertion-mounted packages have leads that extend through the PWB. As the name implies, surface-mount packages are joined to a single surface and may be leadless or leaded. Most modern electronics use surface mountings due to their ease of assembly, higher input/output counts, and reduced land area. Insertion-mount technology, however, is still used extensively in consumer electronics and mixed technology boards are also being produced. Figures B.1 and B.2 show examples of typical surface-mount and insertion-mount interconnects, respectively.

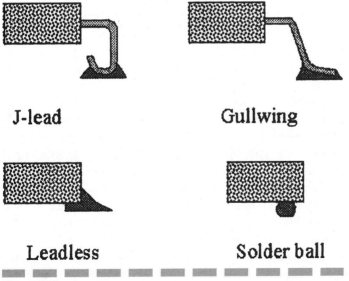

J-lead   **Gullwing**

Leadless   **Solder ball**

Figure B.1
Illustration of typical surface-mount interconnects.

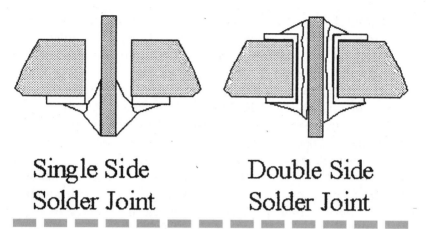

Single Side Solder Joint   Double Side Solder Joint

Figure B.2
Illustration of fatigue-related failure sites for insertion-mount interconnects.

Common insertion-mount packages include the dual in-line package (DIP), the pin grid array (PGA), can packages, and axial discretes (capacitors and resistors). Common surface-mount packages include leadless ceramic chip carriers (LCCC), plastic leaded chip carriers (PLCC), small outline j-leaded (SOJ) packages, small outline gull-wing packages (SOP), thin small outline packages (TSOP), leadless chip capacitors (LCC), and leadless chip resistors (LCR). New surface-mount packages include ball-grid arrays (BGA) and chip-scale packages (CSP). There are currently over 20 different CSP package formats. For surface-mount technology, components are mated to a PWB with solder in a reflow process. For insertion-mount technology, a wave-soldering process is typically employed. In both cases, a tin-lead alloy (solder) is generally used to create the bond.

Eutectic or near eutectic tin-lead (60%Sn 40%Pb) is commonly used in CCAs. A variant of this alloy is a ternary alloy composed of 62%Sn36%Pb2%Ag, which is also used. Investigations into other lead-free systems have been conducted based on the environmental concerns related to lead. Initiatives are in progress to move toward lead-free electronics by the early part of the twenty-first century. These initiatives have spawned a search for replacements for tin-lead solder. Potential candidates include tin-silver, tin-indium-silver, tin-bismuth-silver alloys. At present, there is no clear leader.

Failure analysis of permanent interconnects requires a detailed understanding of the physical behavior of permanent interconnect materials. Under field loading conditions, it is important to understand the viscoplastic nature of solder. Several authors have addressed solder properties and behavior.[6–10] The melting point of eutectic or near eutectic solder is approximately 183°C. With temperature conditions in applications approaching 85°C or higher, the soldered interconnect is approximately one-half of its melting point. As a result, solder will creep at elevated temperatures and the stress in the interconnect will relax. This results in a hysteresis of the solder. Figure B.3 shows the effect of the creep behavior in terms of an idealized stress-strain history. Since electronic products are subjected to temperature excursions, the creep behavior of solder is critical. *Creep* is defined as the time-dependent change in strain under a constant load and consists of an initial high strain rate followed by a steady-state strain rate. The Weertman steady-state creep law can be used to model the creep behavior of solder. For this model the creep strain is defined as

$$\gamma_{cr} = \gamma_0 \sigma^{1/n_c} \exp\left(\frac{-\Delta H}{T}\right) t \qquad (B.3)$$

where $\gamma_0 = 6.82e^{-15}$ $((\text{lb/in}^2)^{1/n_c})/\text{s}$, $1/n_c = 6.28$, and $\Delta H = 8165.2$ for eutectic solder.[8]

In addition to the creep behavior, solder exhibits plastic flow under stress. This behavior can be modeled using the Ramsberg-Osgood relationship, which defines the strain as a function of elastic and plastic components as defined by

$$\varepsilon = \varepsilon_{el} + \varepsilon_{pl} = \frac{\sigma}{E} + \left(\frac{\sigma}{K}\right)^{1/n_p} \qquad (B.4)$$

where σ = the equivalent stress
  E = the modulus of elasticity
  K = the Ramberg-Osgood constant
  n_p = the strain-hardening exponent

For eutectic solder, the yield stress is 4960 lb/in² and $E = 3.62e6$ lb/in², $K = 7025$ lb/in² and $n_p = 0.056$.[8]

The stress-strain history of a solder joint, depicted in Fig. B.3, is directly related to the package, interconnect, and board. For leadless packages, the stress state may reach a plastic yield point during the course of a temperature excursion. If the system is maintained at the elevated temperature for any period of time, relaxation of the stress will occur within the soldered joint. On the downward side of the temperature cycle the reverse occurs. However, there may be little or no relaxation at low temperatures. The outer solid line in Fig. B.3 depicts a hypothetical hysteresis loop for a leadless package. In the case of leaded packages, the stress may not reach the plastic yield state. In this case, relaxation occurs from somewhere below the yield state. As a result, the hysteresis loop is smaller as depicted by the dashed lines. Thus, leaded packages tend to have longer lives than leadless packages under similar temperature cycles.

In order to forecast the overall life of a soldered interconnect, the stress-strain history and damage behavior of the interconnect material must be considered. As discussed previously, the PoF approach is a two-step process that involves evaluating a stress metric and then evaluating the damage metric. Researchers have proposed several stress metrics and damage laws for assessing the fatigue life of the permanent interconnects. These methods are based in stress, strain, and/or dissipative energy. The following paragraphs present several models and approaches for forecasting the fatigue life of soldered interconnects. These damage models include the stress approach (Basquin), the strain approach (Manson-Coffin), and the energy approach.

τ **Hysteresis**

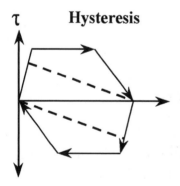

Figure B.3
Idealized hysteresis loop for solder joints.

B.1.2.2 STRESS APPROACH. Fatigue life is often character-ized in terms of the applied stress range. For many engineering materials, the fatigue life exhibits a linear relationship with the applied stress range when presented as a log-log plot. The rela-tionship can be mathematically represented[10] as

$$N_1 \sigma_1^b = C \tag{B.5}$$

where C is the fatigue strength coefficient and b is the fatigue damage exponent. This relationship has a wide range of applica-bility. This model is most relevant for stresses within the elastic limits of the material. The stress metric for this damage model is the average cyclic stress range. Steinberg[11] demonstrated the use of this model to the fatigue life of permanent interconnects under both temperature cycling and vibration load conditions.

The problem with the Basquin damage model is that it does not capture the damage due to inelastic deformation. In general, solder undergoes inelastic deformations making the Basquin model questionable. The Basquin relationship can still be used, but it should be used with caution.

B.1.2.3 STRAIN APPROACH. The strain-based damage models also make use of a power law form but use strain rather than stress as their input metric. The plastic strain life model[12,13] relates fatigue life to the plastic strain range. The damage model is given as

$$N_f = \frac{1}{2} \left(\frac{\Delta\gamma_p}{2\varepsilon_f} \right)^{1/c} \tag{B.6}$$

where c is the fatigue ductility exponent and ε_f is the fatigue ductility coefficient. For eutectic solder, ε_f is approximately equal to 0.625. The *fatigue ducility exponent* has been defined as a function of the mean cyclic temperature and the cycle frequen-cy[6] and is expressed as

$$c = -0.442 - 6.0 \times 10^{-4}T_s + 1.72 \times 10^{-2}\ln(1-f) \tag{B.7}$$

where T_s is the mean cyclic temperature and f is the cycle fre-quency. For this model, the cycle frequency is between 1 and 1000 cycles/day.

A more generalized strain range approach is to combine the elastic strain life (Basquin) and the plastic strain life (Coffin-Manson). This model relates the total strain range (elastic and plastic) to cycles to failure (damage metric) and is given as

$$\frac{\Delta\gamma}{2} = \frac{\Delta\gamma_e}{2} + \frac{\Delta\gamma_p}{2} = \frac{\sigma_f}{E} (2N_f)^b + \varepsilon_f (2N_f)^c \qquad \text{(B.8)}$$

where $\Delta\gamma$ = the total strain range
$\Delta\gamma_e$ = the elastic strain range
$\Delta\gamma_p$ = the plastic strain range
E = the modulus of elasticity
σ_f = stress strength coefficient
b = the fatigue strength exponent
c = the fatigue ductility exponent
ε_f = the fatigue ductility coefficient

The advantage of the generalized Coffin-Manson equation is that it accounts for both the elastic and inelastic behavior of the material. A plot of the generalized Coffin-Manson relationship is presented in Fig. B.4. As can be seen from Fig. B.4, the elastic strain range dominates high-cycle fatigue while the inelastic strain dominates low-cycle fatigue.

B.1.2.4 CRACK PROPAGATION APPROACH. The crack propagation approach is appealing since it represents the physics involved in the interconnect failure. Crack propagation assumes that there is a nucleation period followed by a crack growth phase. Cracks initially occur in the high stress concentration area of a solder joint. For gull-wing leads, the crack generally initiates at the heal where the lead bends to form the foot. Rather than shear loading, the normal (peeling) forces drive the crack formation and growth. Figure B.5 shows a solder-joint fatigue crack for a gull-wing lead. The cracks can be seen as the dark lines through the solder. For BGA interconnects, cracks initiate and grow in the solder ball near the attached substrate. Cracks have been shown to form on the outer edge and move inward as can be seen in Fig. B.6. For leadless interconnects, the crack initiates under the component at the pad-solder interface and grows outward into the sol-

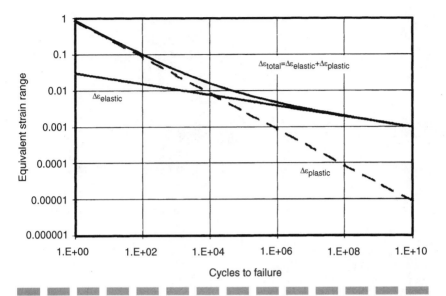

Figure B.4
Generalized Coffin-Manson relationship.

der fillet. Shear forces appear to drive leadless interconnect fatigue failures.

Classical fracture mechanics suggest that the crack initiation and growth is related to a stress intensity factor or the J-integral. Researchers examining soldered interconnects have also related crack initiation related to the strain range or viscoplastic strain energy. In experiments presented by Darveaux,[9] the cycles-to-crack initiation was modeled using the viscoplastic strain energy per cycle, ΔW. In this case, the crack initiation model is given as

$$N_o = C_i \Delta W^{b_i} \qquad (B.9)$$

For 62%Sn36%Pb2%Ag, Darveaux[9] suggests the following material constants: $C_i = 7860$ and $b_i = -1.0$. In this same report, the crack growth was also related to the viscoplastic strain energy per cycle. In this case the crack growth model, defined as the area growth per cycle, was given as

$$\frac{dA}{dN} = C(\Delta W)^b \qquad (B.10)$$

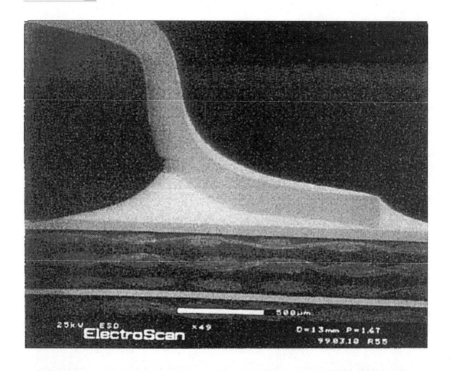

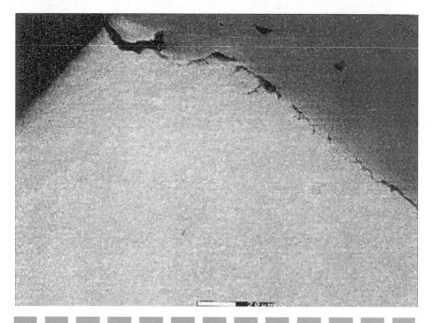

Figure B.5
Fatigue failure of a gull-wing interconnect.

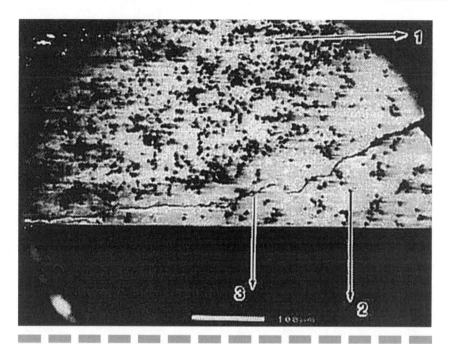

Figure B.6
Broken solder ball interconnect due to fatigue.

For 62%Sn32%Pb2%Ag, $C = 1.0e^9$ and $b = 1.19$. While the crack initiation and growth models provide a more realistic presentation of the failure of soldered interconnects, it presents a significant experimental challenge. The primary drawback stems from the fact that the acquisition of crack-related measurements is extremely difficult for specimens that match the geometries of actual solder joints. The measurement of bulk specimens may not provide relevant data.

B.1.2.5 ENERGY APPROACH. The energy approach relates the failure of the interconnect to the energy dissipated during a stress cycle.[8] In this approach, the number of cycles to failure is proportional to the irreversible energy consumed at an interconnect joint. Like the strain range models, energy models assume the failure is a power law relationship with the hysteresis (cyclic) energy. The most basic model is given as

$$N_f = C\Delta W^m \qquad (B.11)$$

where C and m are material properties and ΔW is the viscoplastic strain energy per stress cycle.

Analogous to the strain range partitioning approach, the need to separate the energies into rate-dependent curves results in the development of the energy partitioning approach.[8] In the energy partitioning approach, the energy dissipated during the stress cycle is partitioned into elastic potential energy, plastic work, and creep work. Each energy component is then related to the cycles to failure. This can be expressed as

$$\text{Energy} = U_e + W_p + W_{cr} = U_o N_{fe}^{b'} + W_{po} N_{fp}^{c'} + W_{co} N_{fc}^{d'} \qquad \text{(B.12)}$$

where U_{eo} = the elastic coefficient
$\qquad b'$ = the elastic exponent
$\qquad W_{po}$ = the plastic exponent
$\qquad c'$ = the plastic exponent
$\qquad W_{co}$ = the creep coefficient
$\qquad d'$ = the creep exponent

A plot of the cyclic strain energy versus cycles to failure is presented in Fig. B.7.

By examining the various fatigue life models, it is clear that the stress history of a solder interconnect is critical for forecasting the interconnect reliability. The stress develops in the solder interconnects as a result of temperature, vibration, and swelling due to moisture absorption. From the literature, it appears clear that temperature followed by vibration is the primary concern. Methods for approximating the stress-strain history, based on temperature and vibration induced loads, are presented in the following sections.

B.1.2.6 TEMPERATURE CYCLING. Temperature cycling of CCAs resulting from power cycling and exposure to uncontrolled environments directly affects useful life. The concern associated with temperature cycling arises from the mismatch of thermal expansion rates between the various materials. The expansion rate of a material is characterized by a coefficient of thermal expansion (CTE), which may be measured experimentally using a thermomechanical analyzer (TMA). A list of CTE common packaging material and packages is provided in Table B.1.

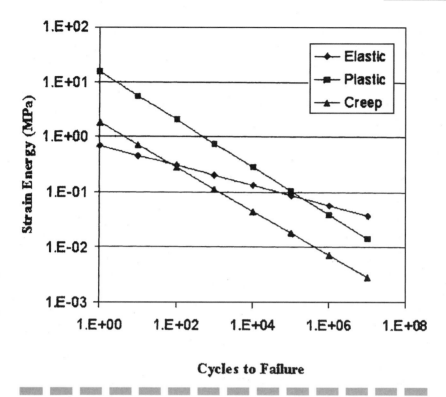

Figure B.7
Strain energy versus fatigue life for 63%Sn37%Pb solder.

For package-to-board interfaces, CTE mismatches may occur globally (package and board) and locally (lead and solder, solder and PWB pad). A depiction of problems related to temperature cycling and CTE mismatch is present Fig. B.8. As a result of the CTE mismatches, the board, package, and interconnect undergo deformation. In general, the interconnect is the weak link and takes up most of the stress and consequently deforms the most. The stress-strain history in the solder joint is an important function of the package and joint geometry, as well as the applied temperature cycle. A discussion of different stress metrics used to characterize thermomechanical-induced fatigue is presented in the following paragraphs.

The stress range resulting from a temperature cycle is a simple metric that can be used to characterize the fatigue of a solder

TABLE B.1

Typical Coefficient of Thermal Expansion (CTE) Values

Material	CTE (ppm/°C)
FR4 (in-plane)	15—20
FR4 (out-of-plane)	54
Plastic packages	20—25
Ceramic packages	5—9
PLCC	22
TSOP	6.7
Kovar	5.9
Solder	23
Polymide fiberglass	12—16

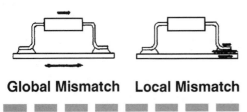

Global Mismatch   Local Mismatch

Figure B.8
Thermomechanical load considerations.

interconnect. The stress range may be obtained by FEA or by the application of basic mechanics of materials principles. This application of basic mechanics of materials principles allows an engineer to obtain a rudimentary understanding of the stress-strain condition within a soldered interconnect. An example of this approach will be discussed.

Figure B.9 shows a surface-mount chip capacitor that is soldered to a PWB. By considering only in-plane forces and ignoring pad-to-solder expansion mismatch, the difference in expansion between the chip and the PWB can be expressed as

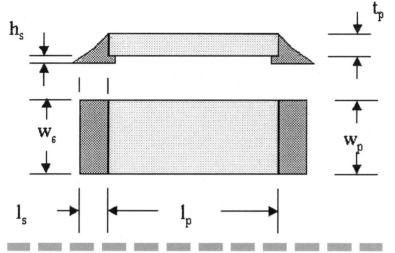

Figure B.9
Leadless chip carrier diagram.

$$\alpha_c L_c \Delta T_c + \frac{P L_c}{E_c A_c} = \alpha_{\text{pwb}} L_c \Delta T_{\text{pwb}} + \frac{P_{\text{pwb}} L_{\text{pwb}}}{E_{\text{pwb}} A_{\text{pwb}}} + \frac{P_{\text{solder}} h_{\text{solder}}}{G_{\text{solder}} A_{\text{solder}}} \quad \text{(B.13)}$$

where $\alpha_{\text{pwb}}, \alpha_c$ = the CTE of the PWB and component

$\Delta T_{\text{pwb}}, \Delta T_c$ = one-half of the temperature cycle range from the high to low temperature dwells for the PWB and component

E_{pwb}, E_c = the modulus of elasticity of the PWB and the component

A_{pwb}, A_c = the cross-sectional area of the PWB and the component

L_{pwb}, L_c = the half lengths of the PWB and the component

G_{solder} = the shear modulus of the solder

h_{solder} = load-bearing height of the solder joint

A_{solder} = the load-bearing area of the solder joint

From this expression, the average force in the interconnect due to a temperature excursion may be determined. A free-body analysis of the chip, the interconnect, and the PWB indicates that

$$P_c = P_{solder} = P_{pwb} = P \qquad (B.14)$$

Based on this simplification, the force, P, acting on the interconnect can be expressed as

$$P = \frac{\alpha_{pwb} L_c \Delta T_{pwb} - \alpha_{chip} L_{chip} \Delta T_{chip}}{\dfrac{L_{pwb}}{E_{pwb} A_{pwb}} + \dfrac{h_{solder}}{G_{solder} A_{solder}} + \dfrac{L_{chip}}{E_{chip} A_{chip}}} \qquad (B.15)$$

From this result, we can see that the shear stress in the solder joint is

$$\tau_{solder} = \frac{P_{solder}}{A_{solder}} \qquad (B.16)$$

A similar approach can be used to approximate the stress for other package styles. The estimated shear stress can be used to forecast the fatigue life by using a Basquin power law relationship for stress versus cycles to failure. As previously discussed, stress is not generally considered a good metric for evaluating the fatigue life of solder. For this approach, calibration factors are needed to obtain good agreement between field failure data and model outputs.

A simple model for estimating the fatigue life of interconnects, based on the cyclic strain range and the cyclic strain energy, was suggested by Engelmaier.[14] In this approach, the strain range is approximated by assuming that the expansion was completely taken up by the solder. In calculating this stress metric, transients and PWB warpage are ignored. The strain range may be approximated as

$$\Delta \gamma = \frac{L_D \Delta (\alpha \Delta T)}{h} \qquad (B.17)$$

where L_D = the half the diagonal length of the package
$\Delta (\alpha \Delta T) = |\alpha_s (T_{s\text{-high}} - T_{s\text{-low}}) - \alpha_c (T_{c\text{-high}} - T_{c\text{-low}})|$

α_s, α_c = the CTEs of the PWB and package

$T_{c\text{-low}}$ = package temperature at the low-temperature dwell

$T_{c\text{-high}}$ = package temperature at the high-temperature dwell

$T_{s\text{-low}}$ = PWB temperature at the low-temperature dwell

$T_{s\text{-high}}$ = PWB temperature at low-temperature dwell

h = the height of the solder joint (or half the solder paste stencil thickness)

To handle leaded packages, strain energy is approximated by considering the stiffness of the lead. The energy is then converted back to a strain range by the introduction of an empirical constant. For a leaded package, the stress metric is defined as

$$\Delta\gamma = \frac{K[L_D \Delta(\alpha\Delta T)]^2}{(200 \text{ lbf/in}^2)Ah} \tag{B.18}$$

where K is the diagonal flexural stiffness of the cornermost lead, A is the effective minimum load-bearing solder-joint area (two-thirds of the lead area projected to the pad), and the other parameters are the same as defined for the leadless case in Eq. (B.17).

The metrics presented in Eqs. (B.17) and (B.18) can be used to estimate the useful life of various packages. A high stress metric normally indicates a low expected life. From inspection, it can be observed that the stress metric is reduced by increasing the solder-joint height, minimizing the temperature expansion differential, or reducing the imposed temperature excursion. For the leaded case [Eq. (B.18)], the stress metric can also be reduced by increasing the load-bearing area of the solder joints, or decreasing the lead stiffness. For the stress metrics defined by Eqs. (B.17) and (B.18), the Manson-Coffin fatigue life relationship is used to forecast fatigue life [see Eq. (B.6)].

Strain energy is a useful stress metric for evaluating the fatigue life of soldered interconnects. While Eq. (B.18) provides an approximation of the strain energy, it does not truly capture the complexity of the stress-strain history in the solder joint. Two major problems are the idealized behavior of the solder joint and the assumed symmetry of the stress-strain history. To truly capture the complexity of the stress-strain history, the stress metric must evaluate the stress state in the soldered interconnect over time.

The time-based evaluation of the stress-strain history in a solder joint can be obtained by deriving the equilibrium equations for a component, interconnect, and board assembly, and then solving for the stress and strain instantaneously with respect to time. This approach was proposed and demonstrated for leadless chip resistors by Jih and Pao[15] and extended to leaded packages by Sundararajan et al.[16] In this approach, a two-dimensional free-body diagram of the package, interconnect, and board is developed. The stress in the assembly is assumed to arise from the in-plane CTE mismatch. A free-body diagram for a lead package is presented in Fig. B.10. The equilibrium analysis yields

$$\gamma + \frac{\tau}{K} + \frac{1}{K_L} = \frac{(\alpha_c - \alpha_p)\Delta T L}{h_s} \tag{B.19}$$

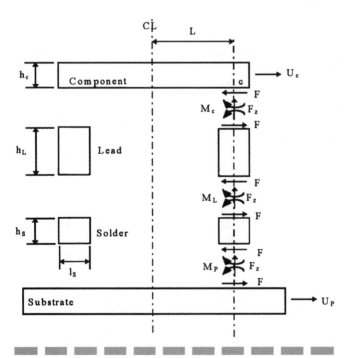

Figure B.10
Free-body diagram of leaded package model.

where γ = the shear strain in the solder
   τ = the shear stress in solder
α_p, α_c = the CTE of the PWB and package
   h_s = the solder-joint height
   L = half the length between the solder interconnect locations
   ΔT = the temperature cycle range
   K = a characteristic stiffness of the surface-mount assembly
   K_L = the stiffness of the interconnect

The stiffness of the assembly is a function of the component and board stiffnesses. The stiffness of the interconnect can be determined by finite element analysis or by application of mechanics of materials principles. Differentiating Eq. (B.19), ignoring plastic yield and using a creep constitutive model for solder to account for creep, yields a first-order, time-dependent, nonlinear differential equation for shear stress in the critical solder joint:

$$\frac{d\tau}{dt} = \left[\frac{1}{K} + \frac{1}{K_L} + \frac{1}{\mu(T)} \right]^{-1} \left[\frac{(\alpha_c - \alpha_p)\, L}{h_s} \frac{dT}{dt} \right.$$

$$\left. + \frac{\tau}{\mu^2(T)} \frac{d\mu}{dT} \frac{dT}{dt} - \gamma_o \exp\left(\frac{-\Delta H}{T} \right) \tau |\tau|^n \right] \qquad (B.20)$$

The stress history in a solder joint can be approximated by solving Eq. (B.19). The strain history can be determined simultaneously by solving

$$\frac{d\gamma}{dt} = \frac{1}{\mu} \left(\frac{\tau_m - \tau_{m-1}}{\Delta t} \right) + \gamma_o \exp\left(\frac{-\Delta H}{T} \right) \tau^n \qquad (B.21)$$

This method was used to analyze a 9.0-mm² 20-pin LCCC subjected to a -55 to $125°C$ temperature cycle. Deriving the assembly and lead constants, then solving Eqs. (B.20) and (B.21) using a differential equation solver yields the stress-strain history of the solder. This information is presented graphically in Fig. B.11 in the form of a hysteresis loop. For comparison purposes, a finite

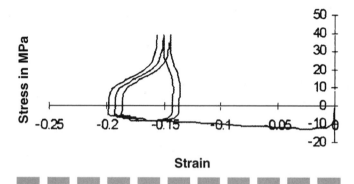

Figure B.11
Stress-strain history using analytic model.

element model of the same assembly was developed and analyzed. The resulting hysteresis plot is presented in Fig. B.12. By examining these two figures, it can be seen that this approach provides a close approximation to finite element results. The two stress metrics derived from these results are potential elastic energy U_o and creep strain energy W_{cr}. These stress metrics are used in an energy partition relationship to forecast fatigue life. Strain range may also be determined from this analysis and used to estimate the value of the Manson-Coffin damage metric.

This section has provided methods for approximating stress-based, strain-based, and energy-based stress metrics. Each approach has its advantages and disadvantages. The major advantages include their representation of the actual physical situation and their speed of solution. A major limitation of these approaches is that they do not capture the more complex stress field that actually exists in the solder interconnect. For instance, none of these methods considers the mismatch of CTE between the solder and pad or solder and the lead. In addition, the stress metrics defined up to this point have assumed a uniform or linear distribution for stress and strain in the solder joint. As a result, calibration factors are often introduced to take into account uncertainties such as stress concentrations, plastic behavior, and manufacturing quality. It is relatively easy to criticize each of the previously defined stress modeling approaches because each has its own individual

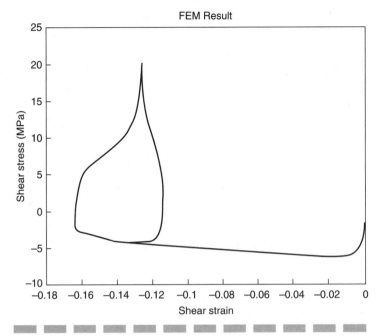

Figure B.12
Finite element analysis stress-strain history.

weakness. However, with proper calibration and rule-based guidance, these models may be successfully used for fatigue life forecasting and design trade-offs.

As discussed previously, the driving mechanisms for solder-joint fatigue failure can be categorized into global and local thermal expansion mismatches. The global expansion mismatch results from the thermal expansion between the component and the substrate. The local thermal expansion mismatch results from the thermal expansion between the lead material and the solder, as well as the solder and the bond pad on the PWB. The relative contributions of the global and local mismatches to solder-joint failure vary significantly. While the global mismatch normally drives the failure, local mismatch for some lead materials (for example, kovar leads) can significantly influence the useful life. Material properties of the component, lead, solder, PWB, and lead compliance[17,18] must be considered in evaluating solder-joint fatigue. Closed-form stress metrics for fatigue analysis of surface-mount solder joints tend to oversimplify the complex stress states that occur in solder joints under cyclic thermal loading conditions.

To capture the complexity of the solder stress state, more sophisticated analytic methods are required. The most common approach is to use a nonlinear finite element analysis to examine the behavior of the solder interconnect under an imposed temperature cycle. Finite element analysis requires a skilled analyst and can be quite time consuming even on today's computer systems. The analysis is time consuming, especially if a design-of-experiments approach is required to determine the optimum design.

To address these problems, Ling and Dasgupta[19] proposed a simplified general-purpose, stress analysis model for both the global and local thermal expansion mismatch in surface-mount solder joints under cyclic thermal loading conditions. In this approach called the multiple domain Rayleigh ritz (MDRR) methodology, the solder joint is mapped in different domains based on the expected stress distribution. Colonies of nested subdomains are introduced where stress gradients are expected. The displacement fields are modeled using general polynomial-based functions. Polynomial displacement fields are superimposed within each nested subdomain in order to model large local displacement gradients. These enhanced displacement fields are carefully chosen in order to maintain continuity between neighboring domains. Continuity between neighboring materials at the lead/solder interface and PWB/solder interface are enforced as constraints using transformation techniques. Unlike conventional finite element modeling where nodal displacements are degrees of freedom, coefficients of the displacement polynomials are the degrees of freedom in MDRR.

An example of the domain mapping for a BGA interconnect is presented in Fig. B.13. The reduction in complexity can be readily seen when compared to a finite element model, shown in Fig. B.14. Since the enhanced displacement fields are only needed at selective regions where stress concentrations are expected, the degrees of freedom contained in a typical MDRR analysis are much less than a finite element model with similar accuracy. The benefits of the MDRR scheme include a simple, quick, and accurate solution of the thermal expansion mismatch problem

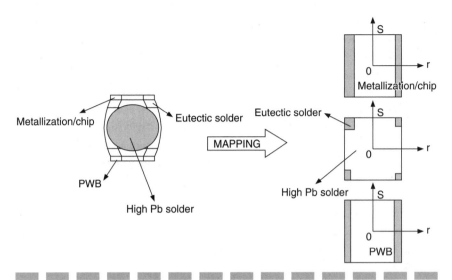

Figure B.13
Domain mapping for a BGA interconnect.

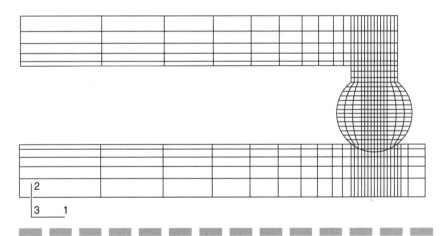

Figure B.14
Finite element model of a BGA.

in surface-mount solder joints, while eliminating the effort of time-consuming finite element model generation.

In the MDRR stress analysis, the elastic-plastic behavior of the solder is considered throughout loading and unloading cycles. The strain energy stored in the solder joint, lead, and

PWB is calculated for each of the temperature increments by minimizing the potential energy. As a result, all the unknown degrees of freedom are solved and the incremental strain and stress are calculated. The strain is further categorized into its elastic, creep, and plastic components and is accumulated throughout the loading and unloading history, accordingly.

In comparing the MDRR approach with finite element modeling, it has both computational advantages and provides similar results. As an example of an MDRR analysis, consider a ceramic BGA with a CTE of 6.4 ppm/°C subjected to a 0 to 100°C temperature cycle. The stress profile of the solder ball at the maximum cyclic temperature calculated by MDRR is presented in Fig. B.15 and by finite element analysis in Fig. B.16. By comparing these figures, it can be seen there is close agreement between the two profiles.

B.1.2.7 VIBRATION-INDUCED STRESS CYCLING. As mentioned previously, fatigue failures may be induced by vibration

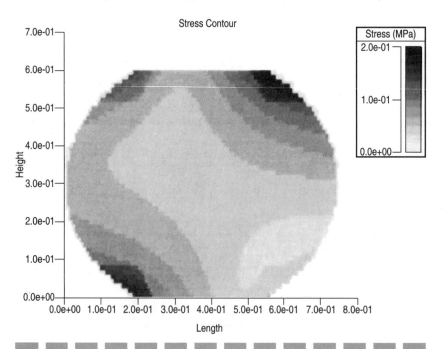

Figure B.15
Stress profile in BGA solder ball as calculated by MDRR analysis.

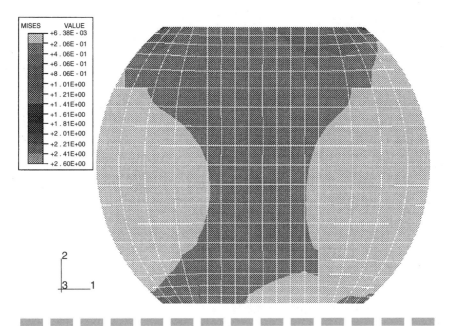

Figure B.16
Stress profile in BGA solder ball as calculated by finite element analysis.

and/or temperature cycling. Vibration loading is characterized by a high number of stress cycles with small amplitudes. Component interconnects are particularly prone to fatigue failure from vibration-induced loads. Failure of materials under vibration-induced loading is typically modeled by the Basquin power law.

For CCAs, test data show that most damage occurs at the fundamental resonant mode where board curvature and stresses are the highest. The stress metric in vibration is the curvature of the PWB, and under simple-simple loading the maximum curvature coincides with the maximum displacement. Based on these assumptions, Steinberg[1] developed a simple stress metric for vibration fatigue of permanent interconnects. In this approximation, the CCA is treated as a simple spring mass system. Based on this assumption, the maximum displacement (curvature) can be expressed as

$$Z = \frac{9.8G}{f_n^2} \tag{B.22}$$

where

$$G = 3G_{rms} = \sqrt{\frac{\pi}{2} Pf_n Q}$$ (B.23)

where P = the power spectral density (PSD) at the natural frequency
f_n = the natural frequency
Q = the transmissibility

The transmissibility is the amount of energy transferred through the supports to the printed wiring assembly. Researchers[1,5] indicate that the transmissibility (amplification) due to vibration is proportional to the square root of the natural frequency, $\sqrt{f_n}$ and is expressed as follows:

$$Q = A \sqrt{f_n}$$ (B.24)

where A = 1/2 for f_n < 100 Hz
A = 1 for 200 < f_n < 300 Hz
A = 2 for f_n > 400 Hz

Amplification increases at lower temperatures, which is countered by the fact that natural frequency tends to drop as the temperatures increase. Low frequencies tend to result in higher curvatures and potentially higher stresses on the component interconnects.

The stress associated with vibration is produced by the maximum board curvature under component and inertia forces. Inertia forces become important for components with large masses, high centers of gravity, and/or weak board-to-package interconnects. For most components, inertia forces are relatively low and curvature-related stresses are dominant in producing failures. An empirical model for determining the maximum displacement at a component was proposed by Steinberg.[1] This model defines the critical displacement as

$$Z_c = \frac{0.00022B}{Chr \sqrt{L}}$$ (B.25)

where B = the PWB length
C = the component factor

h = the board thickness
r = the position factor
L = is the component length

For random vibration, if the displacement is maintained below Z_c, the components are assumed to be able to survive approximately 20 million stress reversals. Based on this information, and using Eqs. (B.22) and (B.25), the fatigue life based on a vibration-induced load can be forecasted by using the Basquin fatigue relationship[1] as follows:

$$N_f = N_r \left(\frac{Z_c}{Z_o} \right)^b \tag{B.26}$$

For electronic components, it has been suggested that $N_r = 20$ million cycles and b is 6.4.[1] This method has been shown to yield good results. A more detailed approach would be to determine the stress based on FEA or the application of mechanics of material principles.

B.1.3   PWB metallization

Metallization on a CCA provides the signal, power, and ground interconnections for components. Reliability concerns for PWB metallization include thermal fatigue of plated through-holes, metal migration in the form of conductive filament formation, dendritic growth, and metal corrosion. These concerns are discussed in this section.

B.1.3.1   PLATED THROUGH-HOLES. The complexity of modern electronics has led to the development of multilayer PWBs, where interconnections of signal and power between various layers is accomplished with plated through-holes (PTHs). A PTH is formed by drilling a hole through the laminated panel from which one or more PWBs will be extracted. The drilled hole is then plated using an electroless plating followed by an electrochemical plating. In general, it is desirable to have 1-mil-thick copper plating and a resistance of below 1 mΩ on the PTH.

The reliability of PTHs has been addressed by several researchers.[11,20–22] The failure mode of the PTH is an increased electrical resistance or opening due to cracking of copper plating. Cracks may be formed by fracture due to overstress or by fatigue when the CCA is subjected to temperature cycling. Overstress fractures are a particular concern during the exposure of the PWB to soldering temperatures above 200°C.[20,21] Temperature cycling during operation can also result in fatigue failure of the PTH.

Stress in the PTH is developed by thermomechanical deformations that occur during temperature excursions. The thermomechanical deformations result from a mismatch of the out-of-plane (z axis) CTE of the PWB and the PTH plating material. Expansion can produce barrel cracking of the PTH plating, separation of the PTH and surface metallization land interface, and separation of PTH and interior metallization interface. Figure B.17 illustrates the effect of temperature cycling on a PTH.

Stress analysis of the PTH under temperature cycling must be conducted to evaluate the likelihood of failure. The stress metrics include the tensile stress in the barrel of the PTH, the stress at the PTH/land interface, and the strain range due to the temperature cycle. The stress and strain data can be determined by finite element analysis or by developing an analytic model.[11,20,23] The generalized Coffin-Manson equation is typically used to estimate the fatigue life of PTHs.

A simple approximation of the stress in the PTH can be obtained by applying basic principles of mechanics of materials. Assuming linear elastic material behavior, the force developed in a PTH subject to a temperature cycle can be defined as

$$P = \frac{\alpha_{pth} L_{pth} \Delta t_{pth} - \alpha_{pwb} L_{pwb} \Delta t_{pwb}}{\dfrac{L_{pth}}{A_{pth} E_{pth}} + \dfrac{L_{pwb}}{A_{pwb} E_{pwb}}} \qquad (B.27)$$

where L_{pth}, L_{pwb} = the thickness of PTH and board
         A_{pth}, A_{pwb} = the area of the board and plating
         E_{pth}, E_{pwb} = the modulus of elasticity of the board and plating
         $\alpha_{pth}, \alpha_{pwb}$ = the CTEs of the PTH and board

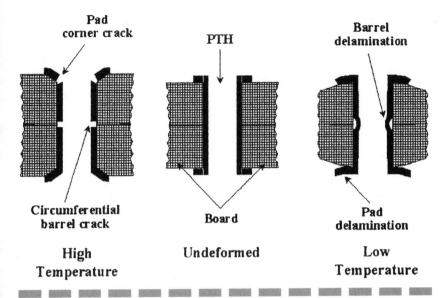

Figure B.17
Effect of temperature cycling on a plated through–hole.

For board area, Steinberg[11] suggests using the length of the PTH versus the hole diameter. The tensile stress in the barrel is then calculated by

$$S_t = \frac{KP}{A_{pth}} \qquad (B.28)$$

where K is the stress concentration factor. The stress is then compared to the ultimate tensile strength or used in a Basquin fatigue relationship.

While the stress approximation provides some quantification of the physical situation, the strain range due to the temperature cycling is more appropriate. Models for approximating the temperature cycle strain range have been developed.[23,24] These models were based on approximating the strain range imposed by the thermomechanical expansion of the PTH and the multilayer PWB. Orien[24] chose to simplify this model by characterizing the multilayer PWB and PTH with effective CTE and modulus of elasticity. Bhandarkar et al.[23] chose to consider the individual plating materials used to create the PTH, as well as the individual layers used to construct the multiple layer PWB. From Bhandarkar et al., the total axial deformation of the PTH is defined as

$$\delta = \frac{\sum_i \alpha_{pth_i} \Delta T_{pth_i} E_{pth} A_{pth_i} + \left\{ \sum_j \alpha_{pwb_j} L_{pwb_j} \Delta T_{pwb_j} \Big/ \sum_j [L_{pwb_j} \big/ (A_{pwb_j} E_{pwb_j})] \right\}}{1 \Big/ \sum_j [L_{pwb_j} / (A_{pwb_j} E_{pwb_j})] + \sum_i A_{pth_i} E_{pth_i} \Big/ L_{pth}}$$

$$(B.29)$$

where $\alpha_{pth}, \alpha_{pwb}$ = the CTE of the plating and PWB layers
A_{pth}, A_{pwb} = the area of the plating and PWB layers
E_{pth}, E_{pwb} = the modulus of elasticity of the plating and PWB layers
L_{pth} = total length of the PTH
L_{pwb} = the thickness of the layers
i = the individual plating material
j = the index for the individual layers

The total axial force in the plating is defined as

$$P = \frac{\delta - \sum_j \alpha_{pwb_j} L_{pwb_j} \Delta T_{pwb_j}}{\sum_j [L_{pwb_j} / (A_{pwb_j} E_{pwb_j})]}$$

$$(B.30)$$

Equations (B.28) and (B.29) are used as stress and strain states in the PWB layers and the through-hole plating. The area used for each layer is assumed to be equal. The value of A_{pwb} is defined as a function of material and geometric parameters listed previously with the addition of a calibration constant. The constant is evaluated by a closed-form equation. The function is introduced to account for pad radius, pad thickness, proximity of other PTHs, board thickness, and the radius of the PTH. The calibration coefficients were determined by a two-level, half factorial numerical design of experiments.

The damage metric for thermal fatigue of the PTHs is cycles to failure and the generalized Coffin-Manson equation was used to calculate the value. The axial strain in the plating was determined by dividing the total deformation by the length of the PTH. A numerical solution algorithm calculates the stress and strain states by incrementally changing the temperature to account for the inelastic plastic behavior of the plating and the property changes that result from exceeding the glass transition point of the board layer material. The results of the algorithm were validated by comparison with experimental data. The experimental data was taken from a comprehensive study of hundreds of PTHs with different configurations on polymide kevlar boards. These configurations were modeled using general-purpose FEA and the developed analytic model. The results of this analysis are presented in Fig. B.18. As can be seen from the figure, the analytic model provides good correlation with experimental results.

B.1.3.2 METALLIZATION CORROSION. Corrosion is an electrochemical reaction that results in the deterioration of a material due to its reaction with the environment. Corrosion is best understood in terms of an electrochemical cell. An idealized electrochemical cell is presented in Fig. B.19. An electro-

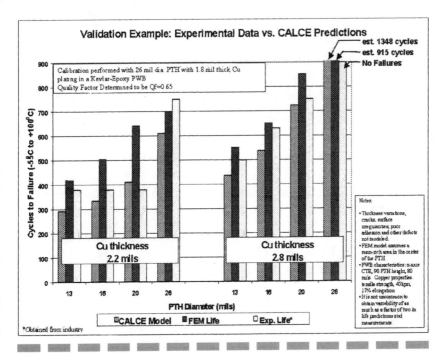

Figure B.18
Comparison of experimental and numerical estimates of PTH fatigue life.

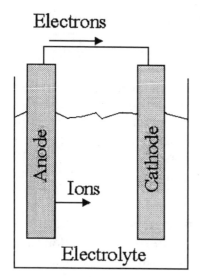

Figure B.19
Idealized electrochemical cell.

chemical circuit consists of two connected pieces of metal in a conducting solution or electrolyte. Features of an electrochemical circuit include:

- *Anode.* The metal is oxidized (ionizes) and corrodes.
- *Cathode.* Under corrosion, the metal ions react with other mobile ions to form gas, liquid, and/or solid by-products.
- *Physical contact.* The anode and the cathode must be in electrical contact with each other.
- *Electrolyte.* The conductive medium that completes the circuit. This medium provides a means for metallic ions to travel from the anode and to accept electrons at the cathode. In many cases, water acts as the electrolyte.

The metal acting as the anode undergoes oxidation and produces metal ions. For CCAs the most prevalent metal is copper and the reaction for copper is defined as

$$Cu \rightarrow Cu^{n+} + ne^- \tag{B.31}$$

Water located at the anode will also react, releasing oxygen and hydrogen ions. The reaction is defined as

$$H_2O \rightarrow \tfrac{1}{2}O_2\,[\uparrow] + 2H^+ + 2e^- \tag{B.32}$$

At the cathode, water will react with mobile electrons as follows:

$$H_2O + e^- \rightarrow \frac{1}{2} H_2[\uparrow] + 2OH^- \qquad (B.33)$$

Metal ions will react with mobile electrons, and for copper, the reaction is as follows:

$$Cu^{n+} + ne^- \rightarrow Cu \qquad (B.34)$$

In electroplating, an imposed voltage is used to drive the reaction, while in corrosion a natural potential is developed. This potential can result from the tendency of metals to give up electrons. Electromotive force (emf), a measure of this tendency, compares the electrode potential of the metal to that of a hydrogen electrode at ambient conditions. A list of electrode potentials for selected materials is provided in Table B.2.

Corrosion may occur as a uniform attack or a galvanic attack. A uniform attack occurs when some regions of metal in an electrolyte are anodic to other regions. Under uniform attack, the anodic and cathodic regions shift over time. Unlike the uniform attack, in a galvanic attack the anodic and cathodic regions remain fixed. Galvanic reactions can occur when dissimilar metals are

TABLE B.2

Electromotive Force (emf)

	Metal	Electrode potential (V)
Anodic	$Ni \rightarrow Ni^{2+} + 2e^-$	-0.25
	$Sn \rightarrow Sn^{2+} + 2e^-$	-0.14
	$Pb \rightarrow Pb^2 + 2e^-$	-0.13
	$H_2 \rightarrow 2H^+ + 2e^-$	0.00
	$Cu \rightarrow Cu^2 + 2e^-$	$+0.34$
	$2H_2O \rightarrow O_2[\uparrow]+4H^+ + 2e^-$	$+1.23$
Cathodic	$Au \rightarrow Au^{3+} + 3e^-$	$+1.5$

SOURCE: Adapted from Ref. 22

connected by an electrolyte. The tendency of a material to have an anodic reaction may be evaluated by examining its position in a galvanic series. A galvanic series lists materials based on their tendency to have an anodic reaction. Materials higher in the list react anodically to materials lower in the list. An abbreviated galvanic series for materials in seawater is presented in Table B.3. As we can see from this table, solder is anodic to copper and will corrode.

In addition to dissimilar materials, galvanic reactions can occur between regions of high and low stress, called *stress corrosion*. Under stress corrosion, high stress areas tend to be anodic to lower stress regions. Corrosion of the high stress area weakens the material and cracks may form.

TABLE B.3

Galvanic Series in Sea Water

Anodic	Magnesium
	Zinc
	Aluminum
	Cast iron
	50%Pb 50%Sn solder
	Lead
	Tin
	Cu-40% Zn brass
	Nickel-based alloys (active)
	Copper
	Cu-30% Ni alloy
	Stainless steel
	Silver
	Gold
Cathodic	Platinum

SOURCE: From Ref. 22

The lack of oxygen also increases the anodic reaction of metals. Because cracks and crevices have lower oxygen concentrations, these areas are particularly susceptible to corrosion. The corrosion of cracks by pitting can further accelerate fatigue failures. This type of galvanic attack is sometimes referred to as *concentration corrosion.*

Corrosion in CCAs can produce failures at multiple sites. Copper, which is used in most PWB designs, is extremely susceptible to corrosion from chlorides and sulfides. In addition, copper oxidizes relatively quickly. Surface corrosion can increase surface film resistance and produce by-products that accelerate wear. The effect of surface corrosion is discussed in more detail in Sec. B.1.4. At present, the effect of corrosion on signal distortion is not well documented. Another corrosion failure risk in PWBs is the reduction of insulation resistance between adjacent conductors. Insulation resistance is lost by the formation of conductive bridges, and corrosion plays a significant role in this area. Failure under this condition may be attributed to dendrite growth or conductive filament formation.

B.1.3.3 CONDUCTIVE FILAMENT FORMATION AND DENDRITIC GROWTH. Conductive filament formation and dendritic growth are related to metallization corrosion. Each can result in the catastrophic failure of CCAs due to the sudden loss of insulation resistance between adjacent conductors. Both conductive filament and dendritic growth involve the migration of metal between adjacent conductors. This phenomenon has been observed to occur in CCAs.[25–27] Dendritic growth is a surface phenomenon where metal migrates between adjacent conductors. Figure B.20 depicts separate paths for dendritic growth.

Conductive filament formation is more insidious because it occurs within the woven laminate. In conductive filament formation, a filament of metal grows along the fiber/resin interface in a PWB between adjacent conductors. This phenomenon has been readily observed during humidity/temperature testing at 85°C/85% relative humidity. Conductive filament paths are depicted in Figs. B.21 through B.23. In both conductive filament formation and dendritic growth, metal migration requires the presence of mobile metal ions, a migration path, and voltage potential. A sign of failure is loss of insulation resistance between adjacent conductors.

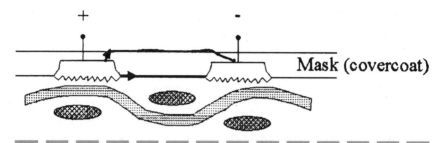

Figure B.20
Dendritic grow between adjacent conductors.

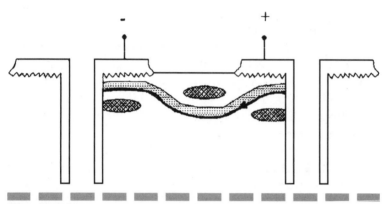

Figure B.21
Conductive filament formation between adjacent PTHs.

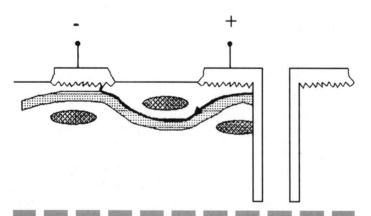

Figure B.22
Conductive filament formation between a trace and an adjacent PTH.

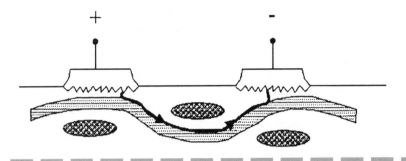

Figure B.23
Conductive filament formation between adjacent traces.

Interfacial failures resulting from dendritic growth are suspected to be caused by hydrolysis. In dendritic growth, a highly localized anodic reaction produces a plumelike structure that extends through the cover coat between the board surface and a region of metallization. The erupted area is the focus of the leakage current. A degradation of the cover coat follows, resulting in exposure of a cathodic conductor. Ultimately, failure results from a short circuit between the adjacent conductors. Surface failure paths may also result from blistering (delamination), which may also open a path for dendritic growth. Dendritic growth and blistering are thought to be results of faulty processing and improper handling.[25]

Studies into conductive filament formation began in the mid-1970s.[25–30] Stress drivers included humidity, temperature, and voltage bias. In initial studies, the migration path was thought to occur due to the breakdown of the interface between the woven fiber and resin matrix due to thermomechanical loading. This hypothesis was supported by studies that showed a 35% reduction in time to failure observed for boards subjected to thermal shock (1 min at 260°C).[28] It was also found that some laminate materials were more resistant to this type of failure than others (for example, triazine showed a marked resistance to this type of failure[30]). Recent studies have indicated that better treatment of the glass fibers and high quality resin can be used to dramatically reduce the likelihood of conductive filament failures.

Another important finding is that the conductive filament formation failures do not require continuous bias.[30] Time to failure of unbiased conductor systems and continuously biased conductor systems did not show significant differences. This suggests that, given the opportunity, conductive filaments can form quickly.

The results of studies into conductive filament formation indicate that failure should be modeled as a two-step process. As a result, the time to failure may be defined as

$$t_f = t_1 + t_2 \qquad \text{(B.35)}$$

where t_1 is the time to establish a migration path and t_2 is the time to form a conductive path between the adjacent conductors. Models for estimating the time to failure have been proposed by several researchers. From analyzing the collected data, Welsher et al.[30] suggested the following:

$$t_f = a\,(H)^b \exp\left(\frac{E_A}{RT}\right) + \frac{dl^2}{V} \qquad \text{(B.36)}$$

where a, b, E_A, and d are material-dependent.

Tests conducted by Rudra and Jennings[31] suggested that the following empirical equation models the data:

$$t_f = \frac{af(1000\,kL)^n}{V^m\,(M - M_t)} \qquad \text{(B.37)}$$

where L = the spacing between adjacent conductors
V = the voltage bias between the conductors
M = the moisture within the laminate
M_t = the moisture threshold
a = the function of the material, geometry, and laminated coating
k = related to the conductor geometry
n = the shape acceleration exponent
m = the voltage acceleration exponent
f = the multilayer connection factor

Failure (that is, growth) was found not to occur for moisture contents below a threshold level. The Rudra and Jennings study included the six-layer laminate test specimens constructed of FR-4, BT (bis-maleimide triazine), and CE (cyanate ester). Each specimen had two test areas that contained various metallization geometeries. In addition, test specimens were not coated, coated with a solder mask, or coated with a postcoat.

While fiber-resin delamination provides a migration path, recent studies[32] indicated that a more insidious path exists in the form of hollow fibers. The topic of hollow fibers will be discussed in Sec. B.2, which addresses manufacturing defects. Dendritic growth and conductive filament formation can be controlled by proper design, material selection, and process control. This supply chain must be considered while addressing conductive filament formation concerns.

B.1.4 Contact and connector failures

Separable interconnects presents another reliability concern[33–35] for CCAs. In most cases, CCAs represent a replaceable item for electronic systems. Separable interconnects provide modularity in electronic systems and allow for field replacement and upgrades. Separable interconnects related to CCAs include sockets for packaged electronic devices, edge connectors for board to motherboard interconnects, and pin-in socket for wiring and peripheral interconnection. The advent of mobile electronics has also lead to the increased use of separable interconnects for peripheral devices.

The function of electrical connectors is to provide electrical interconnections with minimal electrical resistance. Contact resistance is defined by Holms law as

$$R_{contact} = R_{cr} + R_f \tag{B.38}$$

where R_{cr} is the constriction resistance and R_f is the film resistance. Constriction resistance, R_{cr}, is a function of the contact area and the resistivity of the mated contacts. The film resistance, R_f, is a strong function of the contact load. Oxide films can result in up to a 1000 times increase in the contact resistance.[34] For high-reliability applications, a gold surface finish is often used because it provides good electrical characteristics and resistance to corrosion.[33,34] The contact resistance for clean gold surfaces is typically below 1 mΩ.[33]

Connector reliability issues focus around the increase in contact resistance at the metal-to-metal interface. The contact resistance is a function of the materials, the geometries, and the contact area. Changes in the contact resistance can be attributed to a change in contact area and materials due to wear, stress relaxation, and corrosion.

B.1.4.1 WEAR. Wear of connector contacts occurs due to the relative motion between two surfaces. This motion can be caused by mating cycles, vibration, or temperature cycling. The wear can occur by adhesion, abrasion, or brittle fracture. Adhesive wear occurs when there is transfer of metal. In adhesive wear, bonds form between the contacting members that are stronger than the cohesive strength of the metal which leads to transfer and wear as sliding continues. Abrasive wear results from the plowing of the surface by an opposing member that is substantially harder and is generally a result of contact misalignment. Finally, fracture wear occurs with brittle plating, especially when the substrate is easily deformed. The surface develops cracks during sliding which may result in loss of coating. Wear may be combatted by reducing the coefficient of friction between the contact surfaces. Lubricants, such as graphite, may be used to reduce friction and wear.

Wear reduces contact force, which increases contact resistance. Fretting wear can occur when contact surfaces under rapid motion wear through surface plating, which may cause intermittent failures.[11] Cracks form in the contact surfaces and oxides form as a result of local heating. The local oxide regions are nonconductive and result in intermittent opens. This problem is a particular concern for high vibration applications.

Wear may be model based on the applied normal force, relative motion, and material properties. Bayer et al.[36] developed an empirical model for estimating the depth of a scar caused by sliding. In this model the scar depth was given by

$$h = KP^m S^n \tag{B.39}$$

where h is the scar depth; P is the contact force; S is the amount of sliding; and K, m, and n are related to the metallurgy. In their study, Bayer et al.[36] found that n was approximately 0.2 regardless of the metallurgy. The value of m was found to be inversely proportional to the metal thickness. The wear coefficient, K, was found to be proportional to the friction coefficient.

As mentioned previously, electrical contacts are often plated with gold because of its exceptional properties and resistance to corrosion. Due to the insertion/withdrawal cycles the connector undergoes during its life, the gold plating wears away and exposes the underplate or the base metal. When this occurs, the contact resistance increases due to film formation and corrosion on the nonnoble underplate or base metal. Failure is defined as the number of mating/unmating cycles required to penetrate and expose the nickel underlay. An example of wear on a gold-plated PWB contact is depicted in Fig. B.24.

B.1.4.2 STRESS RELAXATION. Stress relaxation of materials used to maintain the contact force in connectors is another area of concern. Contact resistance is a strong function of the normal

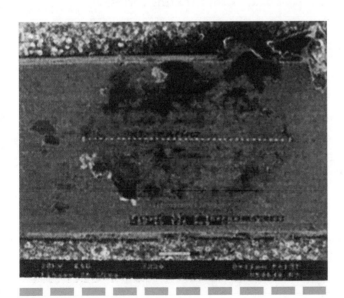

Figure B.24
Wear due to temperature cycling.

force between the metal surfaces of a connector system. Based on the connector design, the contact force may be actively applied or be a part of the connector system. It is common for designs to use a cantilever spring to develop and maintain the normal force. Once mated, the spring system is under a continuous load, which provides the opportunity for materials to yield and thus reduces the contact force. The yielding of a material under a continuous load is termed *stress relaxation*. Copper alloys are used extensively in connectors because of their electrical conductivity, strength, and corrosion resistance.[37] The alloy system and heat treatment can produce variations in each alloy's ability to resist stress relaxation. A graph of several copper alloys is presented in Fig. B25.

In general, a high normal force is desired to provide a low and stable contact resistance. However, the normal force provided by the connector spring decreases with time and temperature due to stress and relaxation in the material of the spring. When stress relaxation occurs, the normal force load decreases, leading to a decrease in the area of the circular spot and an increase in the constriction resistance. In addition, the reduction in force may allow relative motion to occur between the contact surfaces. Thus, stress relaxation may permit fretting corrosion.

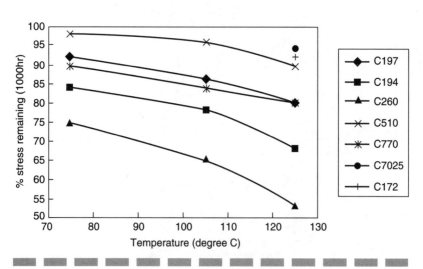

Figure B.25
Stress relaxation of copper alloys.

B.1.4.3 CORROSION.   Usually a copper alloy is used as the base metal for contact pins and springs. They are plated with precious metal to increase corrosion resistance. The gold contact plating typically will be in the range of 0.4 to 1.3 mm thick. However, despite having noble metal plating, corrosion is still a problem due to wear or the porosity in the plating. Porosity is a function of the plating process used, and for a given process the porosity is directly proportional to the thickness of the noble metal plating. When the plating is thicker, the porosity is lower. Porosity exposes the nonnoble underplate or base metal to the environment, which leads to its corrosion. These corrosion products creep out of the pores and spread over the noble metal plating. The corrosion process has been discussed previously in this section.

Corrosion affects contact resistance in several ways. One problem is that corrosion can attack at the periphery of any or all of the contact spots. This causes the constriction resistance to increase as the contact spot area and the spot distributions are reduced in size. Another related problem is that nonconductive corrosion products could form in the spaces between the contact spots. Corrosion products can produce intermittent failures when a contact spot comes in contact with a corrosion product, and then returns to its original position. Alternatively, the contact might ride up on the corrosion product and remain there. Finally, when the contact spot moves back, it might carry the corrosion product back with it. Increased contact resistance is a problem with corrosion products.

B.1.5   Summary of CCA Failure Mechanisms and Models

In this section, common CCA failure mechanisms and models have been discussed. To completely address reliability, the impact of manufacturing and assembly must be understood. Manufacturing and assembly may introduce defects which could reduce the life of the CCA. Even a perfectly designed CCA can be unreliable if manufacturing defects cause early life failure. The next section discusses manufacturing and assembly defects.

B.2   Reliability Issues Related to Manufacturing and Assembly Defects in CCAs

The manufacture and assembly process for PWBs and CCAs can represent several reliability concerns. The concern related to manufacturing and assembly is the introduction of defects. The implication of these irregularities to long-term service reliability is crucial. Most defects fall into three major categories: intrinsic material defects (including improper material properties and cracks), improper geometry (including incorrect size, incorrect thickness), and improper location (including misregistration errors).[38] The presence of defects can increase the susceptibility of a CCA to potential failure mechanisms. Defect magnitudes can be addressed in terms of their effects on time to failure or on performance parameters. While the basic geometry, material, and operation parameters determine a nominal design, the worst-case defect magnitudes determine the defect-related reliability risk. While some defects exist in all CCAs within the limits of the statistical process control, the allowable defect magnitudes can be derived based on the life-cycle stress profile and mission life requirements.

Efficient application of screens can detect defects introduced by the manufacturing and assembly processes. Although it should be noted, some defects may not be screened. Defect types and locations are needed to design effective product screens. CALCE Electronics Products and System Center has a detailed on-line resource[38] for defects related to the assembly of PWBs. Cost-effective screens are determined by assessing the effects of the worst-case defect magnitudes during accelerated tests. The effectiveness of a screen is directly related to the defect for which the screen is designed. Broad or improperly applied screens may not catch intended defects and may reduce the useful life of the CCA. Some screens induce damage to the CCA, including coupon testing and stress screening. Coupon testing allows for destructive testing of samples subjected to the same fabrication process. Stress screens typically subject the CCA to a defined amount of temperature cycling and vibration. Although frequently used, stress screens are costly and have questionable effectiveness. Visual, x-ray, and infrared inspection offer a non-damage-inducing evaluation process.

To understand defects, it is important to understand the fabrication process of a PWB. A PWB is generally constructed by laminating dielectric cores together. The dielectric cores typically consist of an encased woven glass fabric which is then encased in an epoxy resin and sandwiched between two copper foils. To save materials and time, the dielectric cores come in standard rectangular sizes (typically 18 by 24 in) and are sometimes referred to as "panels." As a result, multiple PWBs are typically laid out as a panel. A circuit is formed by removing unwanted copper on the panels by subjecting the treated panels to an etching path. For PWBs with multiple layers, panels are stacked together with a resin prepreg inserted between the individual panels with a copper foil placed above and below the prepreg at the top and bottom of the stack. This stack is sometimes referred to as a "blank." Multiple blanks separated by aluminum are put together to form a book, and the layers are laminated in an evacuated elevated temperature press. Once laminated, the blank is drilled, processed, and etched. The individual PWBs are then extracted from the processed blank in a depanelizing process. After further postprocessing, the PWBs are ready for the assembly process.

Under good process control, PWBs and assemblies rarely exhibit production-related failures.[39] When failures occur, they can generally be traced to defects or deficiencies in materials or the manufacturing process. Throughout this section, defects related to PWB fabrication and the production of PWB assemblies will be discussed as they relate to different steps in the manufacturing and assembly process. The defects related to CCAs can be categorized by physical structures that occur in circuit card assemblies. These categories include PWB laminates, plated through-holes (PTH), and soldered interconnects. Defects related to each of these structures are discussed in the following sections.

B.2.1 PWB laminates

Reliability concerns related to laminate typically are related to loss of insulation resistance, reduction of mechanical strength, and dimensional stability. Defects occurring in PWB laminates include

- Delamination
- Hollow fibers
- Laminate voids
- Plating slivers

These defects are discussed in this section.

In terms of PWBs, *delamination* is the separation of dissimilar materials that are bonded together to form the board. Delamination can occur between the resin and fiber bundles or the resin and conductive metallization. Delamination can often be observed as a change in color in the delaminated area. Figure B.26 shows a cross section of the delamination of the resin from the fiber under magnification. The delamination can be seen as shadow images around fibers at the bottom of the fiber bundle. Delamination of the fiber bundle and resin may occur due to the starvation of epoxy in the glass cloth layers, contaminants present during the coating process, and/or incomplete resin curing. The CTE of epoxy resin is approximately 12 times as large as the CTE of the glass fibers. As a

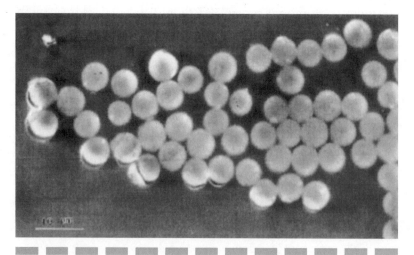

Figure B.26
Delamination of the resin and the fiber. (Courtesy of CALCE EPSC.)

result, epoxy resin/glass fiber separation may arise from mechanical stresses due to the mismatch in the properties of epoxy resin and glass fibers. Temperature extremes encountered under reflow soldering present a particular concern. While generally lower, temperature cycling during usage may also be a concern. In addition, the difference in moisture absorption of glass fibers and epoxy resin can cause interfacial stresses. Metallization delamination can result from the board fabrication process due to excessive pressure and temperature or insufficient bonding of the laminate resin. An example of delamination between plated metal and epoxy fiber core is depicted in Fig. B.27.

Measling and crazing are two forms of delamination where the glass fibers separate from the resin at the weave intersection. Both measling and crazing appear as white spots or "crosses" below the surface of the base material and may appear locally or over a large area. Measling differs from crazing in that it is thermally induced, rather than mechanically induced (crazing). Other factors that influence measling and crazing include moisture in the laminate and improper curing of the laminate.

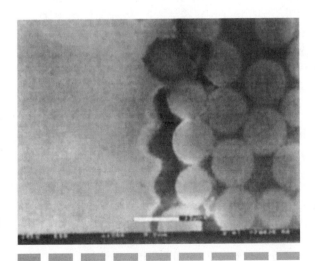

Figure B.27
Delamination between the resin and copper. (Courtesy of CALCE EPSC.)

Depending on the application, there are different board-acceptance criteria for measling and crazing. Ordinarily, measling and crazing are considered to be merely cosmetic defects with no documented board failures due to either measling or crazing. Crazing is generally considered unacceptable, because the almost continuous epoxy separation decreases dielectric strength. Some studies consider measling unacceptable in high humidity usage, because of the possibility for growth of metallic dendrites through the insulator, seriously reducing the dielectric strength. Gross delamination is cause for rejection, because if the resultant cavities fill with moisture or electrolyte, reduced insulation resistance and/or mass-soldering blowholes can occur. Delamination also provides an opportunity for conductive filament formation (CFF).

As discussed in Sec. B.1.3, hollow fibers may also provide an opportunity for conductive filament formation to occur. Hollow fibers are assumed to arise from the fiber fabrication process. This fabrication process consists of four fundamental steps: conversion of raw materials into molten glass, fiber drawing, fabric weaving, and resin coating. In the drawing process, molten glass is fed through a metal plate with tiny holes (bushing) and cooled to form thin glass fibers. Hollow fibers are suspected to form during the glass formation process when gas bubbles become trapped in the molten glass as it is drawn through the bushing. Gas bubbles may arise from trapped air or impurities in the raw materials used to form the glass. If the trapped gas bubbles do not cause the fiber to break, the entrapped bubbles form capillaries in the fiber.

Researchers at CALCE have described a process for identifying hollow fibers by examining the exposed bare glass bundle matrix of a laminate.[40,41] In this process, a sample is excised from the laminate and the surrounding resin is burnt off the sample. The edges of all sides of the test specimens are sealed to prevent wicking (capillary action of a fluid into a hollow fiber), and placed in refracting oil overnight. The oil is selected to have a refractive index that matches the glass fiber. Light, which is directed onto the sample, travels freely until it hits a hollow fiber (air), and the change in the refractive index at the fiber/air boundary will be partially reflected. The use of a microscope with a camera attachment to identify the hollow fibers is recommended, despite the fact that hollow fibers are visible to the naked eye.

The presence of hollow fibers is a particular concern in the formation of fine-line printed circuit boards, laminated multi-chip modules, and plastic ball-grid arrays used in high reliability applications. In these applications, even a short hollow fiber can connect two conductive elements. Since the presence of hollow fibers is likely to increase the chance of conductive filament formation failure, the use of materials with any hollow fibers is not recommended. Further reading on hollow fibers can be found in the references.[42–46]

Laminate voids are defects that occur in multilayer PWB lamination and appear as voids or separations within the laminate material. Generally resulting from thermal stress heat contraction of the epoxy, voids may be adjacent to inner-layer copper foil or a PTH barrel. Void formation may arise from improper curing or flow of the resin, air entrapment, insufficient prepreg, and resin outgasing. Voids may be detected through visual inspection when the defect is gross or very close to the surface of the board. A vertical cross section of the PTH is frequently required to reveal these defects.

Plating slivers are etching defects occurring when unsupported overhang on the surface of the PWB fractures and results in small metallic particles or slivers. An excessive amount of either etch undercutting or edge overhang increases the probability of slivering during subsequent board assembly procedures and during use. Most PWB specifications require that all tin/lead–plated boards, based on epoxy glass, be reflowed to eliminate the tin/lead overhang along the conductor edges or pad rims. In general, there should be no evidence of overhang after etching.

The probability of slivering can be determined by transverse sectioning through representative circuit traces. These traces are examined using an optical microscope to determine the amount of etch undercutting or conductor overhang present. It is also possible to determine the extent of conductor overhang or etch undercutting by comparing the conductor width and the master line width on the 1/1 master pattern at the same location on the printed wiring diagram. Examples of techniques used to determine the amount of undercutting or overhang on PWB circuit traces can be found in the *Electronic Materials Handbook*.[47] Plating slivers are considered to be major defects because they can bridge the dielectric between traces and result in short circuits or dielectric breakdown.

B.2.2 PTH defects

Plated through-holes (PTHs) provide electrical interconnectivity between selected layers of CCAs consisting of more than a single layer. PTHs are normally created by a drilling process. During the drilling process, the blanks are pinned in stacks on tooling plates and drilled on multispindled drill machines. The computerized drill program determines hole locations, and automatically changes drill sizes when a drill size completes its path. To ensure complete drilling and proper alignment with the inner layer pads, the holes may be examined through x-ray and visual inspection. After these inspections, any burrs or resin smear covering inner layer connects are removed. This process slightly roughens the hole and allows for subsequent plating. The hole is plated by copper deposition under an electroless process. The thickness of the electroless layer is typically 80 to 100 millionths of an inch thick. After a number of processing steps, the barrel of the hole is electroplated with copper to a thickness of 1 mil. This is generally followed by plating of either tin or tin/lead. Potential defects in this process resulting from the drilling and desmearing process[48] include

- Etchback problems
- Misregistration
- Nail heading
- Plating folds and nodules
- Plating cracks at the PTHs
- Plating voids and thickness variation
- Resin smear
- Wicking

These defects are discussed in this section.

A desmearing process is used to clean the hole and remove resin that has covered interior metallization resulting from the drilling operation. The desmearing process can result in etchback or the removal of the laminate resin and the woven glass fabric that is used to reinforce the resin from the unplated hole. The desmearing procedure can cause the edges of the internal copper foil to project into the hole. Normally the plating of the hole makes contact with three surfaces of the internal foil: the top of the foil, the vertical edge, and the bottom surface. The proper etchback can increase mechanical strength and ensure good interconnection between the inner layer foil and the copper wall plating.

Excessive etchback of base material can result in unacceptable, nonuniform plating in the hole which does not meet the specified requirements. Inadequate etchback can result from improper curing of the laminate, hardening of the resin smear, insufficient agitation of the desmearing bath, improper bath temperature, and imbalance of chemicals in the desmearing bath. Reliability concerns with etchback include cracking of metallization and exposure of metal in the PTH.

Misregistration is the maximum amount of variation between the centerlines of all terminal pads within one PTH and occurs during board fabrication. It is directly attributed to problems involved with the production of the artwork and their materials, the setup procedures during lamination, or with the dimensional instability of the laminate used. Improper or careless registration of the artwork, improper assembly of blanks, loss of dimensional stability, and worn or deformed holes in the artwork may result in misregistration. Misregistration may be detected by x-ray inspection of the PWB. The amount of misregistration is determined by calculating the difference between centerlines of terminal pads that are shifted to extreme positions. The consequences of misregistration can be very serious, because violation of the minimum clearance could induce a possible short.

Folds and nodules in the plating of a through-hole present a quality control concern. These structures can be caused by rough or improper drilling, loose fibers, and/or insufficient cleaning. Incomplete curing of the resin may also cause folds and nodes. Nodules alone may not be cause for rejection, but they can cause other problems. For example, when a pin gauge is inserted into the holes to verify hole sizes, the nodules can be broken loose, creating voids. Additionally, insertion-mount interconnects may not allow the detection of voids caused by the dislodging of nodules. Later, outgasing of the laminate may occur during soldering. Figure B.28 shows an example of plating folds where the plating, which is shown as a shade in this image, is very uneven and has numerous bulges.

Nail heading is another indication of poor drilling. Nail heading results in flaring of inner layer metallization and is attributed to inadequate cutting of the metal by the drill bit. Dull or worn drill bits and improper drill speeds may cause this defect. The mechanical deformation of the metallization causes hardening and increases the risk of cracking. Figure B.29 shows a microsection of a PTH in which the internal pads develop a cross section corresponding to a nail with a conical head. In this image, cracks in the copper are visible. For multilayer boards, nail heading can increase the internal copper to one or one-half times the thickness at the hole edge. Nail heading is usually not serious in itself, but in extreme cases, internal pads on multilayer boards may be ripped out by the drill.

In addition to nail heading, resin smearing is often found on the internal PTH structure. Resin smear is formed when the heat generated by drilling of a PTH causes the resin of the board to flow and then harden on top of the exposed copper layers. Resin smear may be caused by improper drill speeds and/or incompletely cured laminates. Smearing presents a relia-

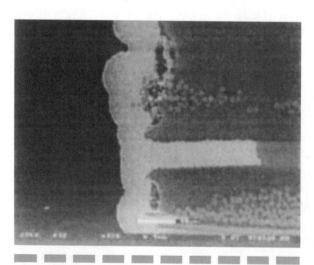

Figure B.28
Example of plating folds. (Courtesy of CALCE EPSC.)

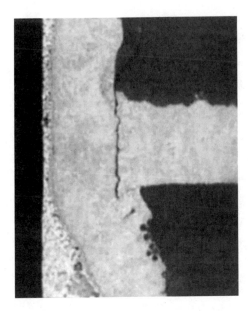

Figure B.29
Example of nail heading. (Courtesy of CALCE EPSC.)

bility concern related to plating. Resin smearing and copper oxide formation may cause interconnection separation and/or marginal interconnections on the inner layer copper foil that contacts the plated copper. Partial and marginal interconnects may fail or be weakened further by the soldering process. Resin smearing can be revealed during microsectioning a suspect PTH. The vertical and horizontal microsectioning of test coupons may be conducted to assess the potential for inner layer disconnections due to resin smearing. If any irregularities are detected between the land and the copper plating, it is likely that a problem exists. However, if there is no visible spacing between the plating and the land, the interconnection is acceptable. The impacts of resin smear are very serious on multilayer boards, because it prevents a reliable plating connection between the inner layers and the hole wall. If there is a separation at the interconnection, the PWB will not function electrically, or it may exhibit an intermittent electrical "open." Any interconnection separation that is detected could result in a failure. In Figure B.30, the vertical cross section shows a resin smear separating the layer foil from the plating.

Cracking plated copper at the knee of the PTH or within the PTH barrel can compromise the reliability of a CCA. Since the

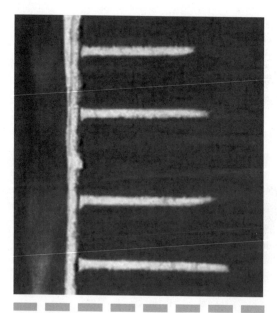

Figure B.30
Example of resin smear and nail heading. (Courtesy of CALCE EPSC.)

CTE of the copper plating and the resin system in the PWBs can differ by a factor of 13, stress is exerted on the plated copper in the plated through-holes in the z axis. Models for approximating this stress were presented in Sec. B.1. Stress cracks can result from inadequate ductility and poor bonding between inner layer metallization and the electroplated copper. Cracks, which support the development of open circuits during the wave soldering operation, can readily be seen in the microsection after the board has been subjected to thermal stress testing. Figure B.31 shows examples of cracks in the PTH plating.

Voids and variation in plating thickness are an additional concern with PTHs. Voids and plating thickness variation can result from lack of proper agitation in the electroless deposition line, insufficient current or time in the plating bath, and/or the presence of contaminants or air bubbles in the hole during the plating process. While the presence of small voids does not necessarily present a reliability problem, voids can represent stress concentration areas. In addition, an isolated area with a thick-

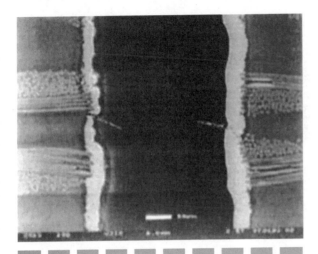

Figure B.31
Cross section of a via with plating cracks. (Courtesy of
CALCE EPSC.)

ness of 0.8 mil on a typical minimum thickness of 1 mil of electrolytic copper may be considered a reliability risk. The presence of excessive plating voids may cause open circuits and
roughness in thin plating, resulting in outgasing of the laminate during the soldering operation. The impacts of insufficient copper plating thickness can be serious if a specific
amount of metal is required to carry the functioning current.
Furthermore, the difference in z expansion between the epoxy
and the copper may cause the thin plating to crack. Figure B.32
shows severe variations in copper plating thickness, resulting in
open circuits.

The extension of copper from a PTH along the fibers generated
by manufacturing defects is referred to as *wicking*. Wicking can
occur along delaminated regions of a fiber bundle and may be
caused by etchback or improper control of chemical cleaning.
Wicking can be serious if it extends sufficiently to deter the dielectric strength or cause internal resistance breakdown between
PTHs. Further, wicking provides a convenient starting point for
conductive filament formation testing criteria on final fabricated
board and inner layer testing. Figure B.33 shows an example of
wicking of copper into the fiber bundle.

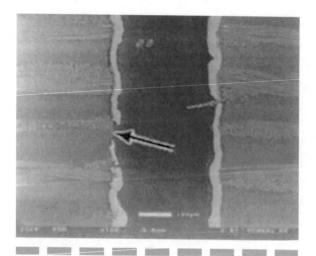

Figure B.32
Example of plating thickness variation. (Courtesy of CALCE EPSC.)

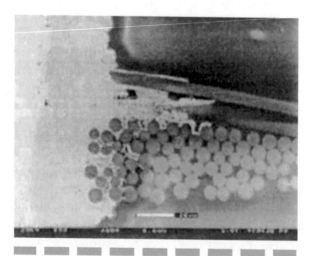

Figure B.33
Example of copper wicking into the fiber bundle. (Courtesy of CALCE EPSC.)

B.2.3   Solder joint defects

As discussed in Sec. B.1, cyclic forces caused by CTE mismatches between the package and the board material and vibration are known to produce wearout failures in soldered interconnects. Defects reducing the joint strength, therefore, are of particular concern and are often related to the "solderability" of the bonding surfaces (for example, PWB pad, component lead, or component pad). Solderability (that is, the ability of leads to form a strong solder joint) is dependent on the wetting angle of the lead/solder joint pair, the lead geometry (for example, shape and coplanarity), the solder temperature, the solder process, and the metallurgy of the bonding surfaces. A more detailed discussion of the soldering process may be found in Ref. 49. Oxides and corrosion on bonding surfaces, intermetallic consumption of bonding surface coatings, and impurities and particulate contamination on the bonding surfaces can reduce solder strength. Corrosion creates soldering problems since solder only forms a strong bond with pure metal or alloys—not a corrosive residue or oxide layer. Adequate cleaning of the leads is required if corrosion products are present on the lead. Improper treatment can result in poor solder-joint quality. Defects related to forming solder joints include [39,47,50,51]:

- Cold solder joints
- Dewetting
- Excessive intermetallics
- Nonwetting
- Solder balling
- Solder bridging
- Solder voids
- Tombstoning

Defects related to soldering are discussed in the remainder of this section.

Cold solder joints are poorly formed joints that often possess a grayish or porous appearance. Cold joints are caused by an improper reflow process resulting from insufficient heat, inadequate cleaning, or excessive impurities in the solder. Cold joints are mechanically weak and produce early failure due to over-stress and fatigue.

Dewetting is a lack of adhesion between solder and the board or package (lead) metallization. Dewetting is caused by a pulling back of the solder on the surface into irregular mounts, high reflow temperatures, and prolonged dwell during the reflow process.[47] These mounds can vary from barely noticeable to having metal exposed between them. Dewetting involves a gas evolution during exposure of the parts being soldered to the molten solder. The gas results from thermally degraded organic material or the release of water from inorganic material. Water vapor at soldering temperature is highly oxidizing and may cause oxidation of the molten solder film surface or subsurface interface. Gas evolution may also be seen where exposed intermetallics exist.

Intermetallics on the lead, which may form between the lead base and the finish, can degrade solderability by consuming the solderable element of the lead coating. Hot-dipping of leads and other high temperature failure accelerators will speed up this phenomenon. An especially susceptible lead base-solder combination is a non-nickel undercoated copper alloy base with a tin or tin-lead solder. Copper-tin intermetallics are not solderable. In addition, the amount of solderable tin is reduced because of the chemical reaction, thereby reducing the solder plating thickness. Nickel undercoatings reduce this intermetallic formation. In addition to the soldering process, intermetallics can occur as a result of prolonged exposure to high temperatures. Intermetallics can reduce the strength of the interconnect and may produce voids in the solder. The gross effect of intermetallic growth is depicted in Fig. B.34. In this case, intermetallics resulted in voiding due to the application of a high tin solder.

Misaligned solder joints are inaccurate positioning of the package interconnects (lead/or bond pads) over the board bond pads where the solder joint forms. Misaligned solder joints can occur due to insufficient tack in the solder paste, poor lead formation, and improper package placement. Solder paste tack is affected by temperature and humidity. Reliability issues with respect to misalignment are related to weak joints which can produce early failures due to fatigue and overstress.

Nonwetting is the case where there is no adhesion between all or part of an interconnect surface and can result from the pres-

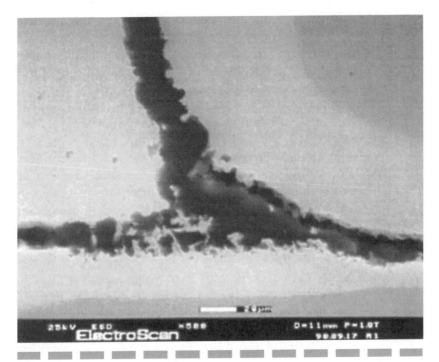

Figure B.34
Solder separation due to intermetallic formation. (Courtesy of CALCE EPSC.)

ence of a physical barrier between the base metal and the solder or a temperature that is too low to allow the solder to wet metallurgically.[47] There is no intermetallic bond between the solder and the base metal in nonwetting. In addition, a surface deposit caused by epoxy or paint on the lead may prevent melting or alloying. An oxidized coating of sufficient thickness may also prevent the melting or alloying. The effective solder area is reduced and also introduces film defects that can act as strong stress risers, which could turn into fracture-initiation sites. The heel area of the joint is most susceptible to cracking because of nonwetting.

Solder balling is the formation of extraneous small solder balls due to lack of cohesion of the solder during the joint formation process. The presence of oxides on the soldering surfaces and/or inadequate or excessive drying of solder paste can cause solder balling. Solder balling can produce short circuits due to undesired bridging.

Solder bridging is the formation of unintended interconnections between adjacent pads and/or leads. Excessive solder paste and inadequate solder masking can cause bridging. Bridging produces short-circuit failures by creating unintended conductive paths between adjacent conductors.

Solder voids are the holes and recesses that occur in solder joints. Voids can be caused during the solder-joint formation by inadequate reflow time, gas evolution from process contaminants, improper solder paste compositions, and interconnect metallization. Voids can also occur because of the growth of intermetallics due to material aging. Failure concerns related to voids are reduced strength, increased stress concentration, and reduced fatigue life.

Starved solder joints are joints that have a lower-than-normal volume of solder between intended interconnection terminals. Starved solder joints can be caused by component lift, insufficient coplanarity of component, insufficient paste, and improper preheating. Gross failure can be observed as open joints. Reliability issues are related to reduced mechanical strength and reduced fatigue life. Figure B.35 depicts a starved joint resulting from the component lift.

Tombstoning is the flipping of a package onto one side during the reflow process. Tombstoning can occur as a result of uneven

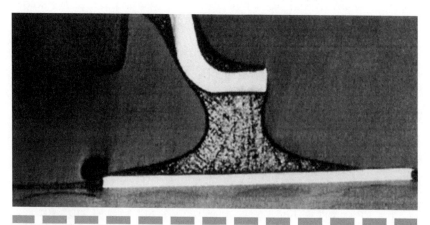

Figure B.35
Starved solder joint due to component lift. (Courtesy of CALCE EPSC.)

past deposition and improper preheating of the assembly. Tombstoning may result in a gross disconnect for electronic terminations on a package. An illustration of tombstoning is presented in Fig. B.36.

Icicling is another defect related to soldering that may degrade the reliability of a PWA. *Icicles* are excessive solder in vias or through-holes that form a point when they cool. Dross and contamination are the two main causes of icicles. In wave soldering, contaminants can cause defects in the solder. The contaminants are the result of the dissolution of plating materials, such as zinc and cadmium, from the surrounding hardware. *Dross* is the formation of oxides in the solder. Dross increases the surface tension in the solder. Icicles can cause solder bridging between pads and lead to board shorts.

B.2.4 Summary

In this section, defects related to the manufacture and assembly process for PWB assemblies have been presented. Defects may result in low yield or cause potential reliability concerns in field product. Although defects are generally low, designers and analysts should have a general understanding of the type and locations of defects. This understanding is particularly important when developing a screening process for PWB assemblies. The selection of qualified manufacturers and review of standard production procedures can aid engineers in reducing and eliminating defects.

Figure B.36
Illustration of tombstoning.

References

1. D. S. Steinberg, *Vibration Analysis for Electronic Equipment*, John Wiley, New York, 1991.
2. M. Pecht, L. Nguyen, and E. Hakim, *Plastic Encapsulated Microelectronics*, John Wiley, New York, 1994.
3. P. R. Engel, T. Corbett, and W. Baerg, "A New Failure Mechanism of Bond Pad Corrosion in Plastic Encapsulated ICs under Temperature, Humidity, and Bias Stress," *Proceedings of the 33rd Electronic Components Conference*, 1986, pp. 127–131.
4. *CalcePWA Explanation and Validations, A Guide for CALCE Software Users*, CALCE EPRC, University of Maryland, College Park, Md., 1996.
5. J. Sloan, *Design and Packaging of Electronic Equipment*, Van Nostrand Reinhold, New York, 1985.
6. W. Engelmaier, "Fatigue Life of Leadless Chip Carriers Solder Joints during Power Cycling," *IEEE Transactions CHMT*, vol. CHMT-6, September 1983, pp. 232–237.
7. H. Solomon. "The Solder Joint Fatigue Life Acceleration Factor," *Transactions, ASME, JEP,* vol. 113, June 1991, pp. 186–190.
8. A. Dasgupta et al., "Solder Creep-Fatigue Analysis by an Energy-Partitioning Approach," *Transactions,* ASME JEP, vol. 114, June 1992, pp. 152–160.
9. R. Darveaux and K. Banerji, "Constitutive Relations for Tin-Based Solder Joints," *IEEE Transactions on CHMT,* vol. 15, no. 6, December 1992, pp. 1013–1024.
10. O. H. Basquin, "The Exponential Law of Endurance Tests," *ASTM Proceedings*, vol. 10, 1910, pp. 625–630.
11. D. Steinberg, *Cooling Techniques for Electronic Equipment*, 2d ed., John Wiley, New York, 1991.
12. L. F. Coffin, Jr., "A Study of the Effects of Cyclic Thermal Stresses on a Ductile Metal," *Transactions,* ASME, vol. 76, 1954, pp. 931–950.
13. S. S. Manson, "Fatigue: A Complex Subject-Some Simple Approximations," *Experimental Mechanics,* no. 5, 1965, pp. 193–226.
14. W. Engelmaier, "Generic Reliability Figures Of Merit Design Tools For Surface Mount Solder Attachments," *IEEE Transactions CHMT,* vol. 16, no. 1, February 1993, pp. 103–112.
15. E. Jih and Y. Pao, "Evaluation of Design Parameters for Leadless Chip Resistor Solder Joints," *Transactions,* ASME, JEP, vol. 117, June 1995, pp. 94–99.
16. R. Sundarajan et al., "Semi-Analytic Model for Surface Mount Solder Joints," ASME IMECE, 97-WA/EEP-12, Dallas, Tex., November 1997.
17. J-P Clech et al., "A Comprehensive Surface Mount Reliability Model Covering Several Generations of Packaging and Assembly Technologies," *Transactions IEEE CHMT,* vol. 16, no. 8, December 1993, pp. 949–960.
18. A. Dasgupta, S. Ling, and S. Verma, "A Generalized Stress Analysis Model for Fatigue Prediction of Surface Mount Solder Joints," *Advances in Electronic Packaging*, ASME, EEP-vol. 4-2, 1993, pp. 979–986.
19. S. Ling and A. Dasgupta, "A Nonlinear Multi-Domain Stress Analysis Method for Surface-Mount Solder Joints," *Journal of Electronic Packaging,* vol. 118, no. 2, June 1996, pp. 72–79.
20. M. Oien, "Methods for Evaluating Plated-Through-Holes Reliability," *Proceedings of 14th Annual IEEE Reliability Physics Symposium,* 1976, pp. 129–131.
21. D. Barker et al., "Transient Thermal Stress Analysis of Plated Through Holes Subjected to Wave Soldering," *Transactions,* ASME, *Journal of Electronic Packaging,* vol. 113, no. 2, June 1991, pp. 149–155.
22. D. Askeland, *The Science and Engineering of Materials*, PWS Engineering, Boston, 1984.
23. S. Bhandarkar et al., "Influence of Design Variables on Thermomechanical Stress Distributions in Plated Through Hole Structures," *Transactions,* ASME, *Journal of Electronic Packaging,* vol. 114, no. 1, March 1992, pp. 8–13.

24. M. Oien, "A Simple Model for Thermo-Mechanical Deformation of Plated-Through-Holes in Multilayer Printed Wiring Boards," *Proceedings of 14th Annual IEEE Reliability Physics Symposium*, 1976, pp.121–128.
25. R. Delaney and J. Lahti, "Accelerated Life Testing of Flexible Printed Circuits, Part II: Failure Modes in Flexible Printed Circuits Coated with UV-Cured Resins," *Proceedings of the 1976 International Reliability Physics Symposium*, 1976, pp. 114–117.
26. P. Boddy et al., "Accelerated Life Testing of Flexible Printed Circuits, Part I: Test Program and Typical Results," *Proceedings of the 1976 International Reliability Physics Symposium*, 1976, pp. 108–114.
27. Bi-Chu Wu, M. Pecht, and D. Jennings, "Conductive Filament Formation in Printed Wiring Boards," *1992 IEEE/CHMT International Electronics Manufacturing Technology Symposium*, 1992, pp. 74–79.
28. J. Lahti, R. Delaney, and J. Hines, "The Characteristic Wearout Process in Epoxy-Glass Printed Circuits for High Density Electronic Packaging," *Proceedings of the 17th Annual Reliability Physics Symposium*, 1979, pp. 39–43.
29. D. Lando, J. Mitchell, and T. Welsher, "Conductive Anodic Filaments in Reinforced Polymeric Dielectrics," *Proceedings of the 17th Annual Reliability Physics Symposium*, 1979, pp. 51–63.
30. T. Welsher, J. P. Mitchell, and D. J. Lando, "CAF in Composite Printed Circuit Substrates: Characterization, Modeling, and a Resistant Material," *18th Annual Proceedings on Reliability Physics*, 1980, pp. 235–237.
31. B. Rudra and D. Jennings, "Tutorial Failure-Mechanism Models for Conductive-Filament Formation," *IEEE Transactions on Reliability*, vol. 43, no. 3, September 1994, pp. 354–360.
32. A. Shukla et al., "Hollow Fibers in Woven Laminates," *Printed Circuit Fabrication*, vol. 20, no. 1, January 1997, pp. 30–32.
33. R. Martens, M. Osterman, and D. Haislet, "Design Assessment of a Pressure Contact Connector System," *Circuit World*, vol. 23, no. 3, April 1997, pp. 5–9.
34. W. Reyes et al., "Factors Influencing Thin Gold Performance for Separable Connectors," *Transactions on Components, Hybrids, and Manufacturing Technology*, vol. CHMT 4, no. 4, December 1981, pp. 499–508.
35. R. Bayer and Gregory, "An Engineering Approach to Vibration Induced Wear Concerns of Electronic Contact Systems," *ASME, EEP*-vol. 14-1, *Advances in Electronic Packaging*, 1993, pp. 525–536.
36. R. Bayer, E. Hsue, and J. Turner, "A Motion-Induced Sub-Surface Deformation Wear Mechanism," *Wear*, no. 154, 1992, 193–204.
37. R. Mroczkowski, *Electronic Connector Handbook*, McGraw-Hill, New York, 1998.
38. CALCE Web Site, *http://www.calce.umd.edu*, reviewed October 1998.
39. R. Tummla and E. Rymaszeewski, *Microelectronics Packaging Handbook*, Van Nostrand Reinhold, New York, 1989.
40. A. Shukla et al., "Hollow Fibers in PCB, MCM-L and PBGA Laminates May Induce Reliability Degradation," *Circuit World*, vol. 23, no. 2, 1997, pp. 5–6.
41. A. Shukla et al., Hollow Fibers in Woven Laminates," *Printed Circuit Fabrication*, vol. 20, no. 1, January 1997, pp. 30–32.
42. Z. Tian and S. R. Swanson, "The Fracture Behavior of Carbon/Epoxy Laminates Containing Internal Cut Fibers," *Journal of Composite Materials*, vol. 25, no. 7, November 1991, pp. 1427–1444.
43. J. Z. Wang and D. F. Socie, "Failure Strength and Damage Mechanisms of E-Glass /Epoxy Laminates under In-Plane Biaxial Compressive Deformation," *Journal of Composite Materials*, vol. 27, no. 1, 1993, pp. 40–480.
44. A. Shukla, M. Pecht, J. Jordon, K. Rogers, and D. Jennings, *Circuit World*, vol. 23, no. 2, 1997, pp. 5–6.
45. "IPC-CC-110 Guidelines for Selecting Core Constructions for Multilayer Printed Wiring Board Applications," IPC-CC-110, Institute for Interconnecting and Packaging Electronic Circuits, Lincolnwood, IL, January 1994.

46. N. Patel, V. Rohatgi, and L. James Lee, "Micro Scale Flow Behavior and Void Formation Mechanism During Impregnation Through a Unidirectional Stitched Fiberglass Mat," *Polymer Engineering and Science*, vol. 35, no. 10, May 1995, pp. 837–851.
47. *Electronic Materials Handbook*, vol. 1, *Packaging*. ASM International. ASM International, Materials Park, Ohio, 1989.
48. M. Oien, "Methods for Evaluating Plated-Through-Holes Reliability," *Proceedings of 14th Annual IEEE Reliability Physics Symposium*, 1976, pp. 129–131.
49. M. Pecht, *Soldering Processes and Equipment*, John Wiley, New York, 1993.
50. Kester Solder Technical Notes, *http://www.metcal.com/kester/*, reviewed February 1999.
51. D. L. Millard, "Solder Joint Inspection," *Electronic Materials Handbook Packaging*, vol. 1: *Packaging*, ASM International, Materials Park, OH, pp. 735–739.

INDEX

415

ABOUT THE AUTHORS

LEONARD MARKS has over 35 years of aerospace and commercial industry experience in printed circuit assembly design and manufacturing. He was a pioneer in the use of computer-aided systems for circuit board layout, and has implemented and managed CAE/CAD/CAM systems at several major corporations. Mr. Marks' manufacturing background includes installation and operation of state-of-the-art circuit board fabrication, assembly and test facilities at three different sites. Mr. Marks served on the board of directors of the International Microelectronics and Packaging Society, and was on the staff of *Electronic Packaging and Production Magazine* as contributing editor. He is a graduate of the Polytechnic University of New York.

JAMES A. CATERINA has over 28 years of experience designing printed circuit assemblies for several major military and industrial electronics manufacturing companies. He is manager of the printed circuit design group at Northrop Grumman's Illinois facility. Mr. Caterina has contributed to the development of several IPC standards.